BASIC ELECTRONICS E

Digital Systems

Malcolm Plant

Hodder & Stoughton
LONDON SYDNEY AUCKLAND TORONTO

One, two, plenty

Tasmanian method of counting

First published in Great Britain 1976

Second edition 1990

British Library Cataloguing in Publication Data.
Plant, M. (Malcolm), 1936–
Basic electronics
Book E. Digital systems
1. Electronic equipment. For schools
I. Title
621.381

ISBN 0–340–41491–X
ISBN 0–340–41490–1 set

Typeset in Baskerville by Taurus Graphics, Abingdon, Oxon. Printed and bound in Great Britain by Thomson Litho Ltd, East Kilbride for the educational publishing division of Hodder and Stoughton Ltd. Mill Road, Dunton Green, Sevenoaks, Kent, TN13 2YA.

Basic Electronics is published in five parts

Book A Introducing Electronics
ISBN 0 340 41495 2

Book B Resistors, Capacitors and Inductors
ISBN 0 340 41494 4

Book C Diodes and Transistors
ISBN 0 340 41493 6

Book D Analogue Systems
ISBN 0 340 41492 8

Book E Digital Systems
ISBN 0 340 41491 X

It is also available as one complete volume:
ISBN 0 340 41490 1

Note about the author

Malcolm Plant is a Principal Lecturer in the Faculty of Education at Nottingham Polytechnic. He is the author of several books, including *Teach Yourself Electronics* (Hodder & Stoughton 1988). His main professional interests are in astronomy and astrophysics, electronics instrumentation and issues relating to conservation and the environment.

Contents

Summary

The general aim of Book E is to show how integrated circuits (ICs) are linked together to produce complete electronic systems such as counters, clocks and control systems. The techniques for making integrated circuits are described in Chapter 16, but in Books D and E we are largely concerned with understanding how ICs are used rather than how they work.

Chapter 1 of this book describes the basic differences between analogue and digital systems. Whereas analogue sytems respond to and produce continuously changing signals, e.g. temperature, digital systems respond to and produce signals which have just two values, on and off, or high and low. In order to show the values of these on/off signals, seven-segment LED displays and liquid crystal displays are used as described in Chapter 2.

One of the most important uses for digital ICs is in decision-making using logic gates: the basic gates and their functions are described in Chapter 3. Decision-making circuits can be analysed using a type of binary maths called Boolean algebra; an introduction to this important algebra is given in Chapter 4.

The use of flip-flops in binary counters and dividers is described in Chapters 5 and 6. This includes binary-coded decimal (BCD) counters for counting digital signals, and dividers in timing circuits. Chapter 7 deals with the use of ICs for coding and decoding digital signals, e.g. changing a BCD signal into the signals required to operate a seven-segment display.

An important use of digital systems is in handling signals from different sources, and routing them to different destinations — this is called multiplexing and is introduced briefly in Chapter 8. Like analogue devices such as the op amp (Book D), digital ICs can be used for making astables and monostables (Chapter 9).

For those readers who want a deeper understanding of digital systems, Chapters 10 to 13 deal more fully with binary arithmetic, binary codes, Boolean algebra, and flip-flops. Chapter 14 looks briefly at the use of digital circuits for adding and comparing binary numbers, two techniques which form the basis of computer operations. Although Book E does not include the detail of computer systems, the purpose and general function of random-access memory (RAM) and read-only memory (ROM) devices are described in Chapter 15. This section ends with a brief discussion of the way a magnetic bubble memory works, and the properties of an exotic semiconductor called gallium arsenide, a material which is so important to the design of future super-fast computers.

Chapter 17 contains a number of applications for digital ICs, including the design of a combinational lock and a sequential lock, a tachometer, and a music box. Chapter 18 contains a description of seven Project Modules. These are commercial units for the rapid assembly of electronic systems. They can be built from the printed circuit board design provided. You should turn to Chapter 12 of Book A for an introduction to their use.

As with every book of *Basic Electronics*, Book E ends with Answers to Questions in the text, Revision Questions and Revision Answers.

If you want to follow an easier and quicker route through Book E, you should omit the sections marked between the symbols ∇ and $\triangle$ in the left hand margin.

1 Introduction

1.1 Analogue and digital systems

As explained in Book A, Section 3.5, electronic systems are broadly divided into two types, analogue and digital systems, depending on the type of electrical signals they are designed to respond to, act on, or deliver. Figures 1.1 and 1.2 will remind you of the difference between the two types of system.

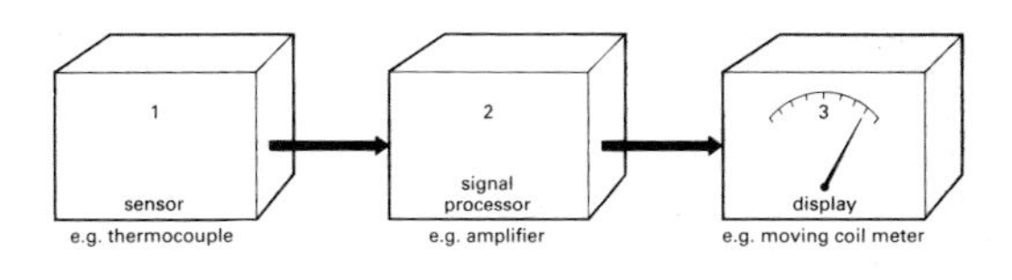

Figure 1.1 An analogue instrumentation system

Figure 1.1 shows the elements of an instrumentation system for measuring temperature. Temperature is an analogue quantity since temperature changes smoothly from one value to the next. In this thermometer, a transducer (black box 1) converts temperature into a smoothly changing electrical analogue of the temperature variation. Black box 2 produces an amplified electrical analogue of the temperature change. Black box 3 then displays the temperature as a smoothly changing analogue reading on a moving coil meter. From start to finish, and including the quantity being measured, the system shown in figure 1.1 is an entirely analogue system.

Figure 1.2 shows the elements of an instrumentation system for counting the number of pills entering a bottle. This number is a digital quantity. A pill either enters the bottle or it doesn't. The number of pills entering the bottle is counted in whole numbers: . . . 10, 11, 12, etc. Digital systems are designed to respond to, act on, or deliver signals which change abruptly from one value to the next. The number of pills is a discontinuous quantity unlike the change of temperature discussed above.

In figure 1.2, the transducer which detects the entry of each pill into the bottle from the pill dispenser is a light source and sensor. The sensor, e.g. a phototransistor, produces an on/off, i.e. digital, electrical signal as each pill momentarily breaks the beam of light. These on/off signals are passed to a counter which operates the digital display showing the number of pills which have entered the bottle. From start to finish, and including the quantity being

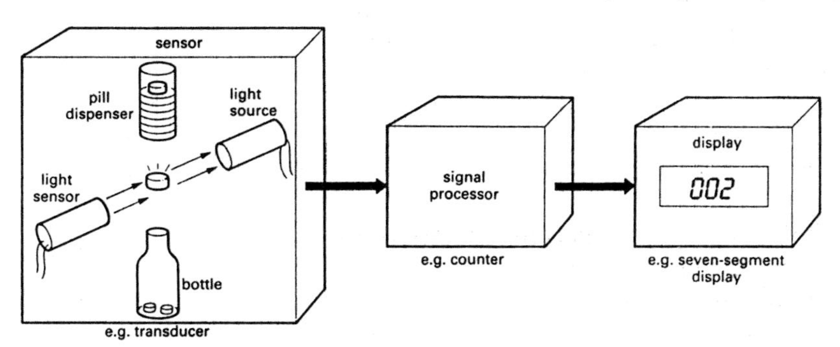

Figure 1.2 A digital instrumentation system

measured, the system shown in figure 1.2 is an entirely digital system.

In Book E, we shall be looking at the design and use of digital electronic systems. Book D describes the design and use of analogue systems.

Questions

1 Is the number of people at a disco an analogue or a digital quantity?
2 While driving, would you prefer to read a digital or an analogue display of road speed?

1.2 Digital inputs and outputs

Figure 1.3 shows that many types of electronic system are made up of three black boxes: input, signal processor, and output — see also Book A, Chapter 3. A digital input to such a system produces signals which are either present (a high signal) or absent (a low signal). Input devices such as switches and square wave generators produce digital signals, and output devices such as counters and relays make use of the digital signals.

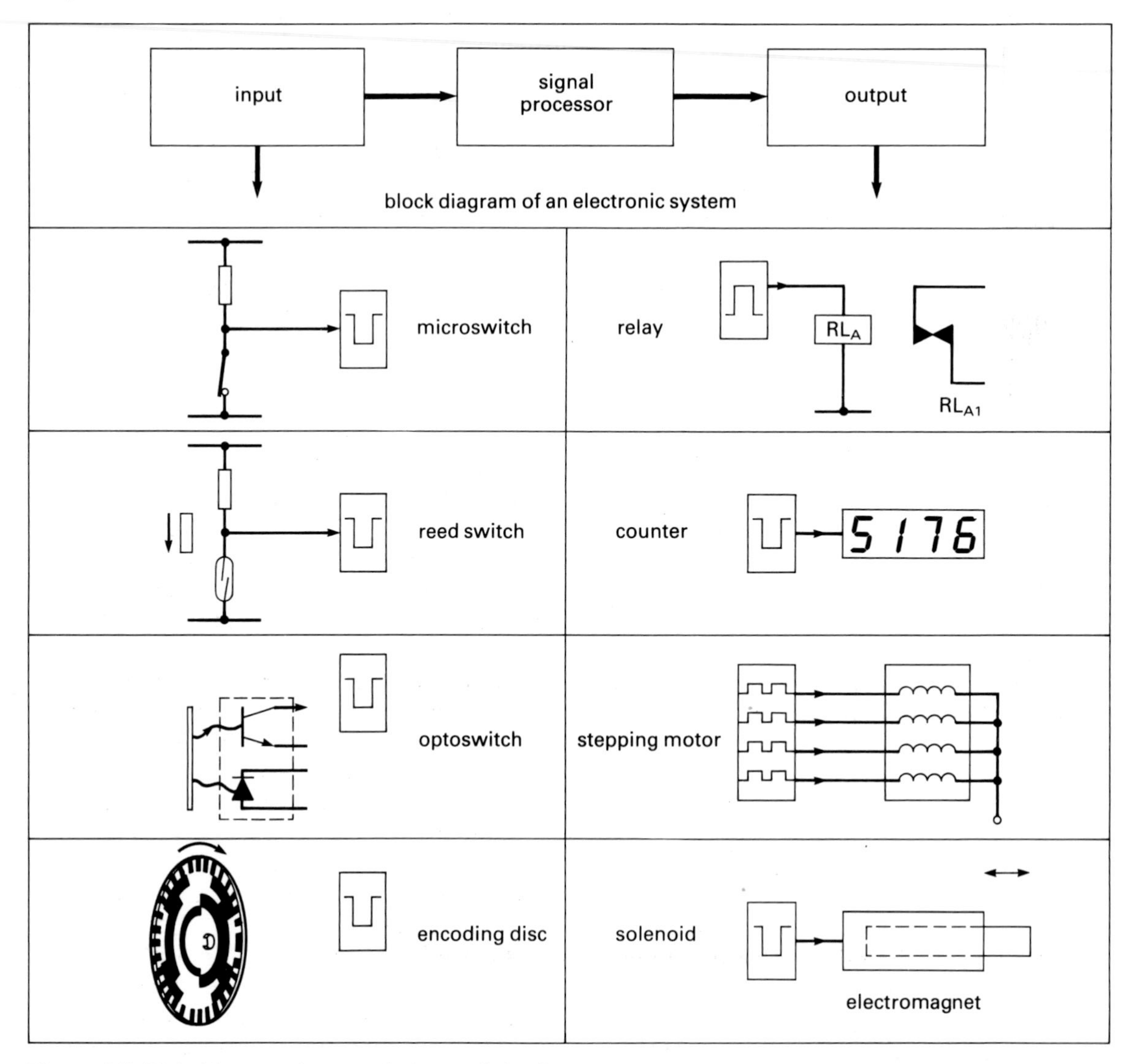

Figure 1.3 Digital input and output devices and circuits

(a)

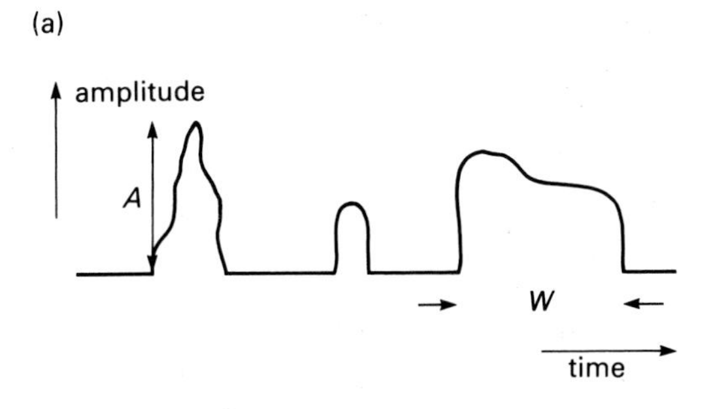

(b)

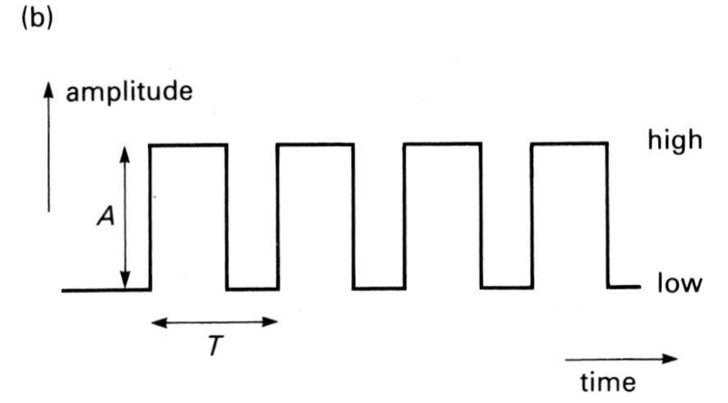

Figure 1.4 Shape of digital signals: (a) irregular series of pulses; (b) regular series of pulses

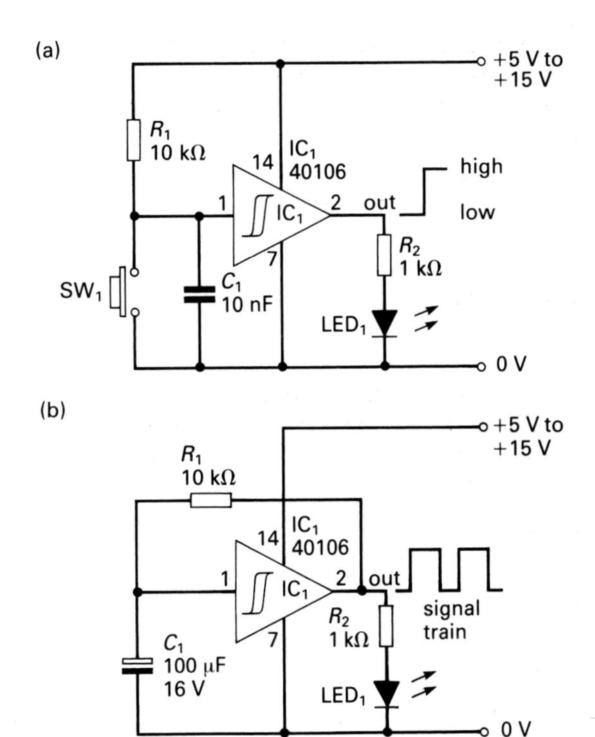

Figure 1.5 Two ways of making digital signals: (a) bounceless switch; (b) electronic 'clock'

A digital electronic signal is made up of one or more pulses as shown in figure 1.4. Each pulse is a short-lived variation of voltage or current, or some other electrical property. It has a pulse width, W, measured in seconds, and a pulse height, A. Pulses such as these would be developed by the switches shown in figure 1.3, e.g. a reed switch which closed and opened for each rotation of a bicycle wheel. The digital signals generated and used in digital electronic circuits are generally square or rectangular. If the pulses occur at regular intervals, they have a period, T seconds, and a frequency, f, equal to $1/T$, just as for regular analogue signals.

1.3 Making digital signals

Figure 1.5(a) shows a practical circuit which produces digital signals every time SW_1 is pressed and released. The light emitting diode, LED_1, shows when the output is high (LED_1 on) or low (LED_1 off). This circuit uses a CMOS integrated circuit, IC_1, a Schmitt trigger type 40106 which is described in Chapter 9. The 40106 contains six Schmitt triggers, only one of which is used here. Pin 1 is the input, and pin 2 the output, of one of these Schmitt triggers. Many mechanical switches suffer from contact bounce, i.e. they produce a series of closely separated pulses as their springy contacts open and close several time. The Schmitt trigger, helped by capacitor C_1, ensures that only one digital signal is produced each time SW_1 is pressed and released. Hence figure 1.4 is called a 'bounceless switch'. Note that the input (pin 1) is pulled low, i.e. to 0 V to make the output go high. This is because the 40106 is an *inverting* Schmitt trigger.

Figure 1.5(b) produces a continuous series of on/off signals — it is an electronic 'clock'. It is based on the same Schmitt trigger as used in the bounceless switch of figure 1.5(a). In this circuit the 40106 is wired as an oscillator which produces a continuous train of clock pulses. With the values of R_1 and C_1 shown, these high/low/high . . . signals switch on and off at a frequency of about 1 Hz. The approximate frequency, f, of the pulses is given by the equation $f = C_1 \times R_1$. The easiest way of changing the frequency is to use a different value for R_1, or change it for a 47 kΩ variable resistor.

1.4 Using digital signals

Figure 1.6(a) shows a practical circuit which accepts digital signals from the bounceless switch or the electronic clock and stores them as a binary number. The value of this number is displayed on the four light emitting diodes, LED_1 to LED_4. This circuit uses a CMOS integrated circuit, IC_1, a binary counter. Pin 15 of IC_1 accepts the digital signals. Pins 2, 14, 11 and 10 are the output pins connected to the light emitting diodes. The binary numbers displayed are from 0000 to 1111 (0 to 15 in decimal). If SW_1 is pressed, all the light emitting diodes go off and the count is reset to zero, i.e. 0000. Chapter 10 discusses binary numbers in more detail.

Figure 1.6(b) shows how a relay can be energised by the digital signals produced by the bounceless switch or the electronic clock. The relay is necessary if high current loads such as motors and lamps are to be switched on and off by the circuits of figure 1.5. Tr_1 is a VMOS transistor (Book C, Chapter 27) which switches on the current required to energise the relay when it receives a high digital signal on its gate (g) terminal. A low digital signal switches off the relay.

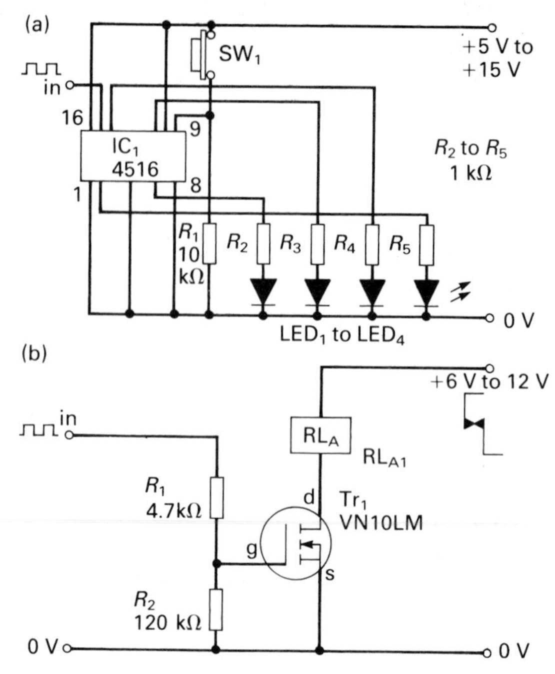

Figure 1.6 Two ways of using digital signals: (a) binary counter; (b) relay driver

2 Digital Displays

2.1 Analogue versus digital displays

The moving coil meter and the bargraph display shown in Book D, figure 4.3, are devices which display information in analogue form; their scales show all possible values of the information between the upper and lower points on the scale. The instrument shown in figure 2.1 has a display of digits which changes by whole numbers, not smoothly from one value to the next as in an analogue display.

Digital displays on instruments, e.g. clocks, watches, point-of-sale terminals in shops, weighing machines, etc., are catching on so fast that they must have some advantages compared with analogue displays. For one thing, digital displays, and the circuits which operate them, are easier to manufacture than analogue displays. Second, the digital readout from a watch can easily be read at a glance since the precise time is shown in hours, minutes and seconds. And digital displays and their circuits are more readily able to sense and measure quantities which are already in digital form. The repetitive rotation of a wheel, the steady beat of the heart, the number of people at a cricket match, are examples of digital quantities.

Question

1 Are there any disadvantages of digital displays?

The increasingly widespread use of digital displays in recent years, has been due to the development of some very clever integrated circuits. These ICs have made the job of processing both analogue and digital quantities, and displaying the result in digital form, much simpler for the manufacturer of digital instruments. Whereas such instruments once required dozens of transistors and ICs, nowadays a single very small chip can perform all the

Figure 2.1 Digital displays in a 'sport computer'

signal processing functions required to display analogue and digital quantities in digital form.

Question

2 Name three electronic devices which you think have been the result of the miniaturisation of electronic circuits.

Figure 2.2 shows an integrated circuit for use in a digital clock. It processes the regular pulses from an oscillator so that time in hours, minutes and seconds can be displayed in digital form. Note that the chip on which the miniature circuit is formed is very much smaller than the package. The plastic package is required to hold the chip and protect it, and it supports the pins needed to connect the chip to an external circuit.

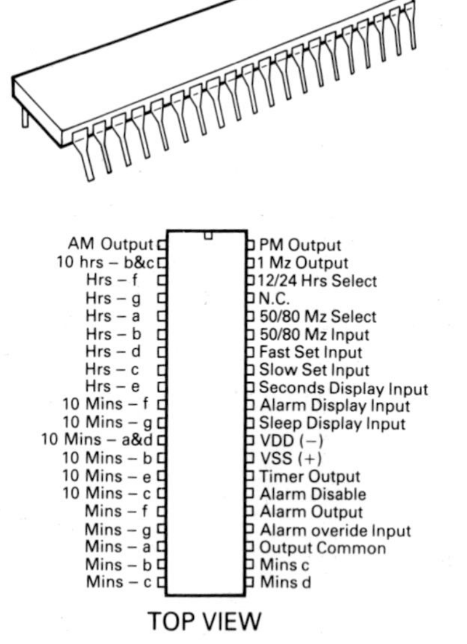

Figure 2.2 A digital integrated circuit for an electronic clock

Question

3 Do you know the names of the two common types of electronic digital display used in clocks, watches and calculators?

2.2 Dot-matrix and seven-segment LED displays

Alphanumeric displays are designed to display numbers, letters and other symbols by illuminating selected dots or bars. The combination of electronics and optics to display information is part of the fast-moving field of electronics called *optoelectronics*.

The general appearance of a dot-matrix display is shown in figure 2.3(a). Letters and numbers are formed by illuminating a matrix of dots. The dot-matrix display has many individual light emitting diodes (LEDs) (see Book C, Chapter 11) which, when lit, appear as dots of light. A typical example is the 4 × 7 LED matrix as shown in figure 2.3(b). A particular LED (e.g. the

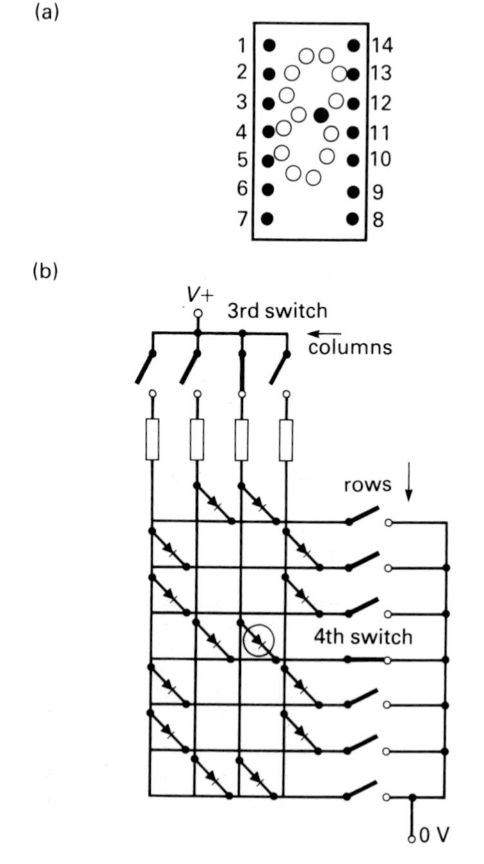

Figure 2.3 A 4 × 7 dot-matrix LED display: (a) appearance; (b) matrix of LEDs

one circled) is lit up by applying a positive voltage to its anode from the 3rd column and grounding its cathode on the 4th row. By applying a voltage to more than one row, and grounding more than one column, it is possible to display any number or letter of the alphabet. In real circuits, the switches are replaced by fast-acting transistor switches.

In the bar-matrix display, one or more bars or segments are lit up, or enhanced in some way, to form symbols. There are two types of bar-matrix display in common use: the seven-segment LED display and the seven-segment liquid crystal display (LCD).

The general appearance of an LED type seven-segment display is shown in figure 2.4(a). Seven bar-shaped segments, a to g, are arranged to display different letters and numbers. Figure 2.4(b) shows how the decimal number 6 is produced by illuminating segments c to g. In a real circuit, the mechanical switches are replaced by transistor switches as explained in Book C, Chapter 18. The decimal number 1 is obtained by illuminating segments b and c, and so on. Only decimal numbers from 0 to 9, and a few special symbols (such as –) and a few alphabetical letters (such as C, c, and F) can be displayed with these seven-segment LED displays. The display has a decimal point (d.p.), another LED, to the right or left of the digit. LED seven-segment displays provide alphanumerics in a variety of colours: red is very common, but orange, green and yellow can be obtained.

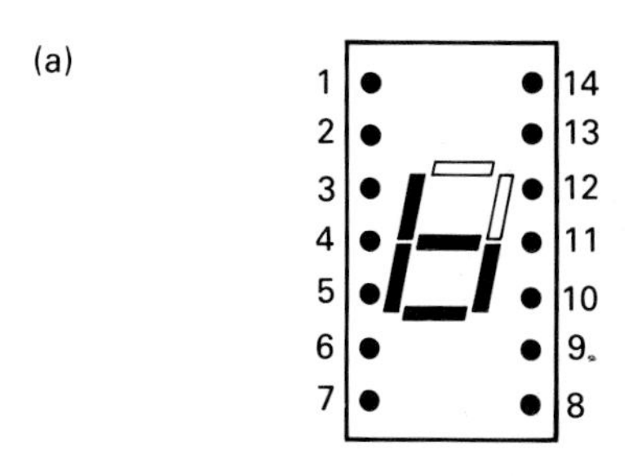

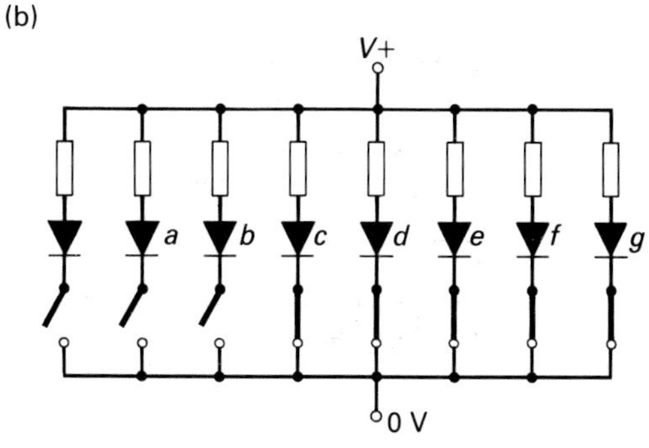

Figure 2.4 A seven-segment LED display: (a) appearance; (b) LEDs in parallel

2.3 The liquid crystal display (LCD)

Whereas the LED actually generates light, the LCD simply controls available light. The LCD has become widely accepted for displaying all types of information, not just seven-segment decimal. For example, the sport computer in figure 2.1 displays an analogue scale of speed above the digital readout. Pocket games, and even some oscilloscope screens and TVs, make use of LCDs.

The LCD relies on the transmission or absorption of light by certain organic carbon crystals which behave as if they were both solid and liquid; i.e. their molecules readily take up a pattern as in a crystal and yet flow as a liquid. In the construction of a common LCD unit shown in figure 2.5, this compound is sandwiched between two closely spaced transparent metal electrodes which are in

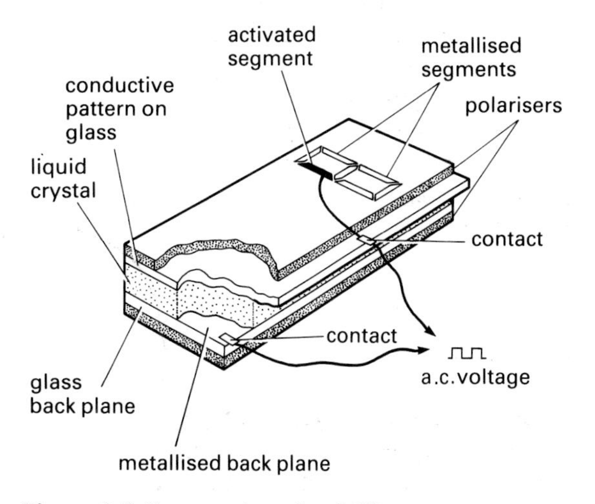

Figure 2.5 Construction of an LCD

the form of a pattern, e.g. a seven-segment digit. When an a.c. voltage is applied across a selected segment, the electric field causes the molecular arrangement in the crystal to change, and the segment shows up as a dark area against a silvery background. A polarising filter on the top and the bottom of this enhances the contrast of black against silver by reducing reflected light.

LCDs that produce frosty white characters on a dark background are also available but much less commonly used. This type is usually the *dynamic scattering LCD*. It uses a different liquid crystal and no polarisers, and it consumes more power than the field-effect type.

The main advantage of the LCD is its lower power consumption and it can be seen in strong sunlight. But it does have the disadvantage that it cannot be seen in the dark, so some watches and clocks are provided with an internal backlight for night time viewing.

2.4 Experiment E1

Testing seven-segment LED displays

There are two types of seven-segment LED display, the *common-anode* type and the *common-cathode* type. In the common-cathode display, the cathodes of all eight LEDs (including the decimal point LED) are connected together internally while the pins give access to their anodes. All eight anodes are connected together inside the common anode display.

Figure 2.6 shows how you can use a breadboard to find out which pins are connected to the segment LEDs of a common-cathode seven-segment display. This display is a DL704 type, but other types have the same pin layout.
Different common-cathode seven-segment displays have different pin connections so make sure you check the 'pin-outs' in the distributer's component catalogue.

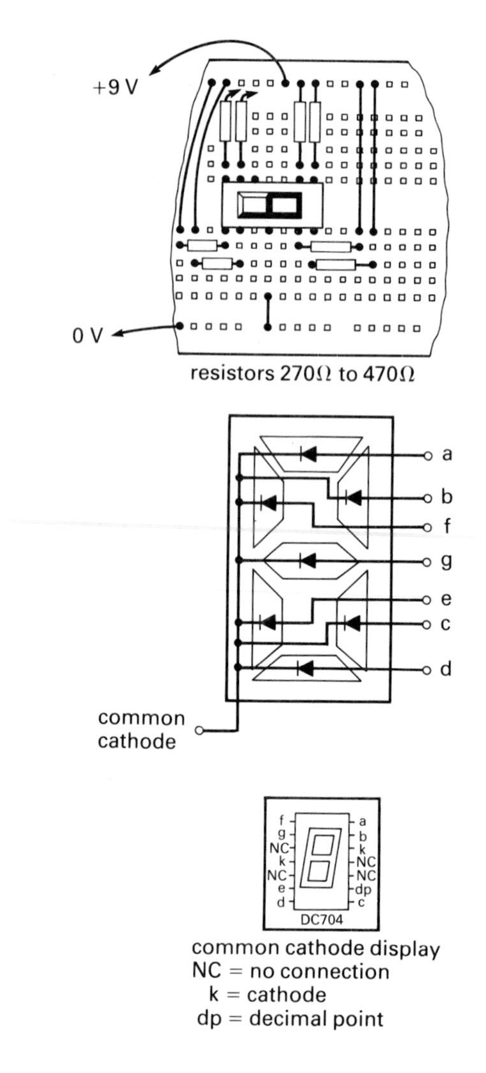

Figure 2.6 Experiment E1: Testing seven-segment LED displays

The anode of each LED in the common-cathode display has to be made positive in order to forward-bias the LED and light it. But remember that the current through the LEDs must be limited by the resistors R_1 to R_7. These should have values in the range 270Ω to 470Ω using a 9 V PP9 battery. Six resistors are shown connected to the positive supply to light segments c to g, but not a and b. Hence '6' is the decimal number displayed. If the other two segments are connected to + 9 V via the wire links, '8' is displayed. See what letters you can display.

3 Logic Gates and Decision-making Circuits

3.1 Introduction

Computers, calculators, control circuits and other equipment that process digital signals are not as magical as many people imagine. Their circuits operate extremely precisely according to strict, logical rules.

The basic building blocks that make up all digital systems are simple little circuits called *gates*. A gate is a decision-making building block which has one output and two or more inputs. It makes decisions based on the inputs it happens to be receiving at the moment.

Figure 3.1(a) shows a simple gate which has two inputs and one output. The inputs are provided by two switches (connected in series). These switches are either open or closed. The output operates a lamp which is either on or off. So this series circuit, which was introduced in Book A, Section 4.4, 'makes the decision' to switch the lamp on when switch A *and* switch B are closed. It is a simple example of an *AND gate*. Similarly, figure 3.1(b) shows two switches connected in parallel. This simple *OR gate* makes the decision to switch the lamp on when switch A *or* switch B is closed.

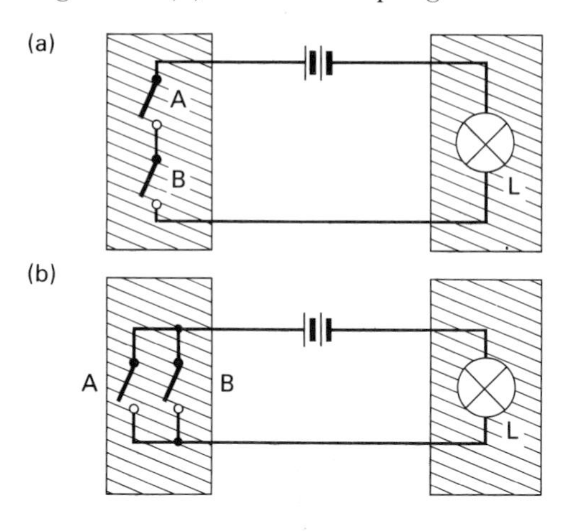

Figure 3.1 AND and OR logic gates using switches (a) AND gate (b) OR gate

Figure 3.2 represents as a block diagram the function of many digital systems. In this sytem, building blocks such as AND and OR gates have an intermediate task of receiving signals from sensors (e.g. keyboard switches and temperature sensors), making decisions on the basis of the information received, and sending an output signal to circuits which provide some action (e.g. switching on a motor). Since these gates are logical in their function, they are sometimes called *logic gates*. And *digital logic* is concerned with the design and use of digital circuits using logic gates.

Now, AND and OR gates are provided in integrated circuit packages. So are the other three basic logic gates: NOT, NAND

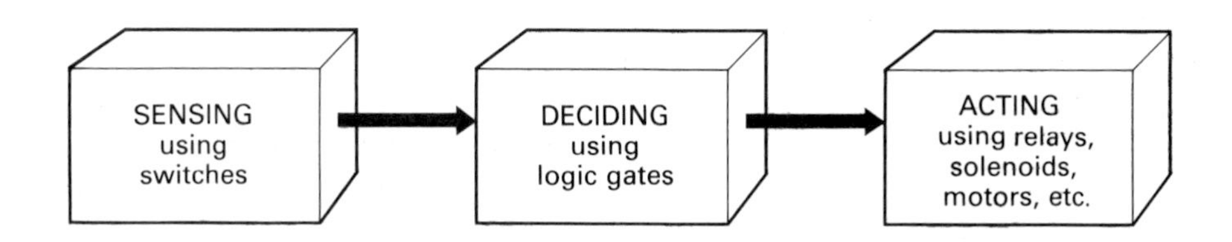

Figure 3.2 A decision-making system

and NOR. There are two main 'families' of digital integrated circuits: transistor-transistor logic (TTL), and complementary metal-oxide semiconductor logic (CMOS). Examples of these packages are given for the logic gates described below. The two families of integrated circuits are compared in Section 3.10. Note we shall use the usual abbreviation 'CMOS' rather than 'CMOSL'.

3.2 Positive and negative logic

In the majority of digital circuits, there are two clearly defined voltages (logic levels) which represent digital signals. These two levels are *logic high* and *logic low*. If the binary number 1 is assigned to logic high, and 0 to logic low, the circuits use *positive logic*: *negative logic* assigns 0 to logic high and 1 to logic low. Positive logic is used in all the circuits in *Basic Electronics*.

In practical circuits using the popular TTL and CMOS logic ICs, the following voltages are used for logic high and logic low:

Figure 3.3

positive logic	TTL at 5 V	CMOS at 10 V
logic high = 1	above 2.4 V	above 9.5 V
logic low = 0	below 0.4 V	below 0.5 V

3.3 The AND gate

The AND gate simply gives an output of logic 1 when both inputs are at logic 1. Thus for a 2-input AND gate, the output is high when both inputs are high. The output is low for all other input combinations. For multiple input AND gates, all inputs have to be high for logic high output. Thus the AND gate is sometimes called an *all-or-nothing gate*.

Figure 3.4 shows three types of circuit symbol used for AND gates. These symbols are discussed in Section 3.12. The truth table gives the state of the output, 1 or 0, for all input combinations — hence the term *combinational logic* for logic systems designed using AND gates — see Section 5.1.

Note the shorthand method of writing the function of an AND gate. 'Input A and input B gives output S' is written A.B = S. The dot between A and B indicates the AND symbol. This equation is a *Boolean* expression and is one of several expressions in Boolean algebra which helps engineers to design decision-making digital logic systems.

Examples of AND gates in integrated circuit packages are shown in figure 3.4. These are quad 2-input AND gates. Note

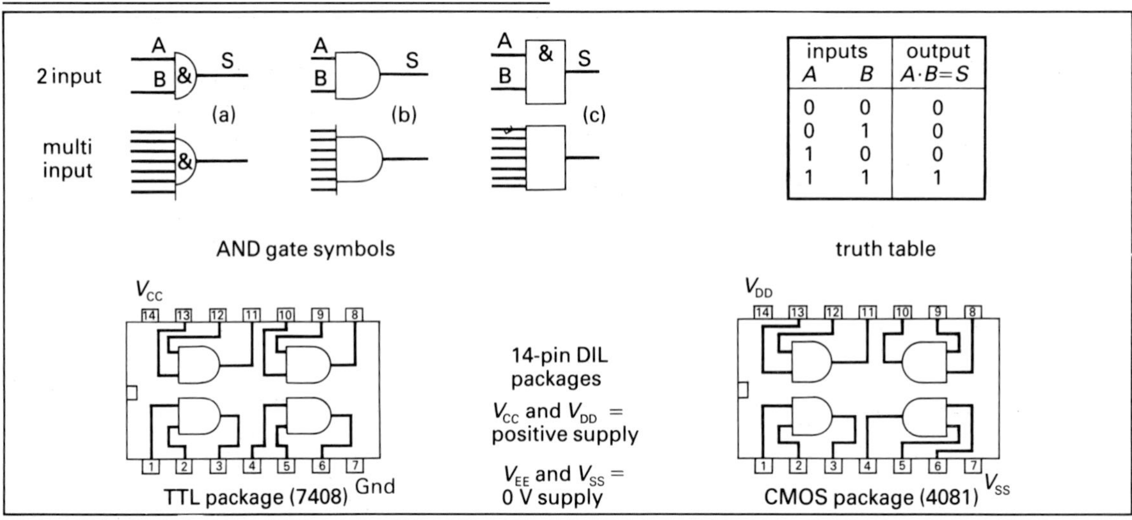

Figure 3.4 AND gate symbols, truth table and packages

that the CMOS gates are 'nose-to-nose' in the package, and the TTL gates are 'nose-to-tail'. Section 3.10 compares TTL and CMOS digital ICs.

Question

1 Write out the truth table for a 3-input AND gate.

3·4 The OR gate

The OR gate simply gives an output of logic 1 when either or both inputs are at logic 1. Thus for a 2-input OR gate, the output is high when either input, or both inputs, are high. The output is low when both inputs are low. For multiple input OR gates, any high input gives a high output. Thus the OR gate is sometimes called an *any-or-all gate*. (See the exclusive-OR gate described in Section 3.8).

Figure 3.5 shows three types of circuit symbol used for OR gates. These symbols are discussed in Section 3.12. The truth table gives the state of the output, 1 or 0, for all input combinations — hence the term *combinational logic* for logic systems designed using OR gates.

Note the shorthand method of writing the function of an OR gate. 'Input A or input B gives output S' is written A + B = S. The plus sign between A and B indicates the OR symbol and is not to be confused with the AND symbol which is a dot, described in Section 3.3. This equation is another Boolean expression.

Examples of OR gates in integrated circuit packages are shown in figure 3.5. These are quad 2-input OR gates. Note that the CMOS gates are 'nose-to-nose' in the package, and the TTL gates are 'nose-to-tail'. Section 3.10 compares TTL and CMOS digital ICs.

Question

1 Write out the truth table for a 4-input OR gate.

3·5 The NOT gate

The NOT gate simply gives an output which is the inverse of the input, i.e. an input of logic 1 gives an output of logic 0, and vice versa. Thus the NOT gate is also called an *inverter*. As shown in figure 3.6, the NOT gate has one input and one output.

Figure 3.6 shows three types of circuit

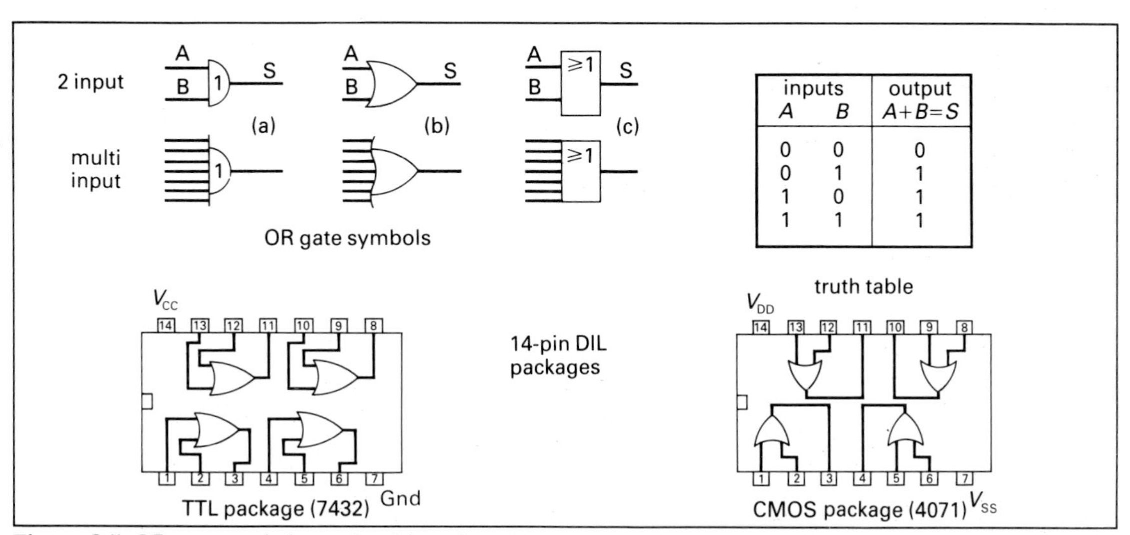

inputs A	inputs B	output A+B=S
0	0	0
0	1	1
1	0	1
1	1	1

Figure 3.5 OR gate symbols, truth table and packages

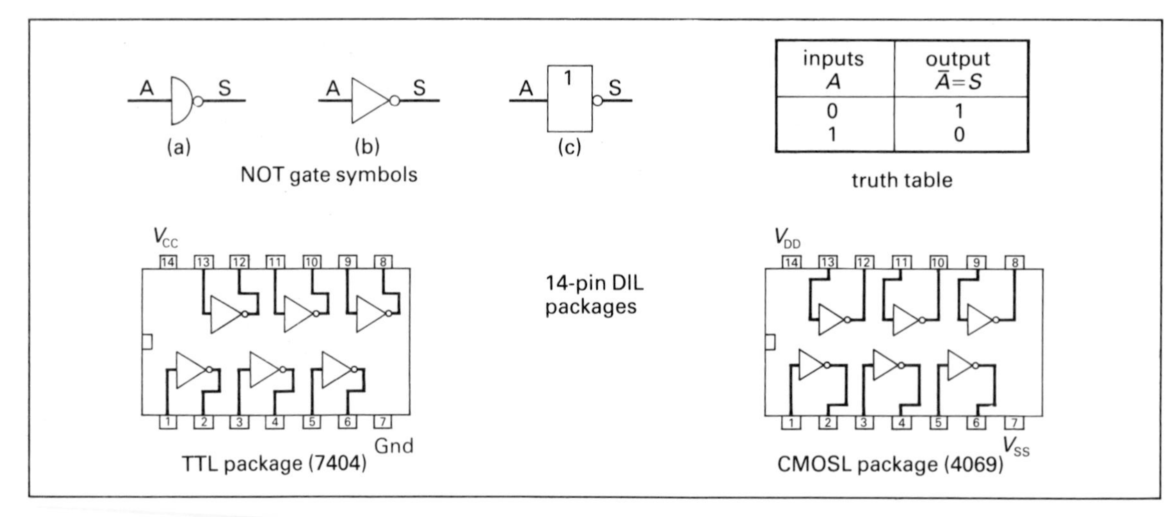

Figure 3.6 NOT gate symbols, truth table and packages

symbol used for NOT gates. These symbols are discussed in Section 3.12. The truth table shows how a logic 1 input gives a logic 0 output and vice versa.

Note the shorthand method of writing the function of a NOT gate. Input A is inverted to give output S and is written $\overline{A} = S$. The bar over the A indicates that it has been inverted, or *complemented*, at the output. This equation is another Boolean expression.

Examples of NOT gates, or inverters, in integrated circuit packages are shown in figure 3.6. These are hex inverters since there are six NOT gates in each 14-pin dual-in-line (DIL) package. Section 3.10 compares TTL and CMOS digital ICs.

3.6 The NAND gate

As you might expect, the NAND (or NOT-AND) gate gives an output of logic 0 only when all of its inputs are at logic 1. Thus for a 2-input NAND gate, the output is low when both inputs are high. The output is high when either or both inputs are low. For multiple input NAND gates, a low on any input gives a high output; the output is low only when all inputs are high.

Figure 3.7 shows three types of circuit symbol used for NAND gates. These symbols are discussed in Section 3.12. The truth table gives the state of the output, 1 or 0, for all input combinations — hence the term *combinational logic* for logic systems designed using NAND gates.

Note the shorthand method of writing the function of a NAND gate. The statement 'not (input A and input B) gives output S' is written $\overline{A.B} = S$. The bar across the top of the AND expression A.B indicates that the output is obtained by inverting the output of the AND gate. This equation is a Boolean expression.

Examples of NAND gates in integrated circuit packages are shown in figure 3.7. These are quad 2-input NAND gates. Note that the CMOS gates are 'nose-to-nose' in the package, and the TTL gates are 'nose-to-tail'. Section 3.10 compares TTL and CMOS digital ICs.

Question

1 Write out the truth table for a 4-input NAND gate.

3.7 The NOR gate

As you might expect, the NOR (or NOT-OR) gate gives an output of logic 0 if any input is at logic 1. Thus for a 2-input NOR gate, the output is low when either or both

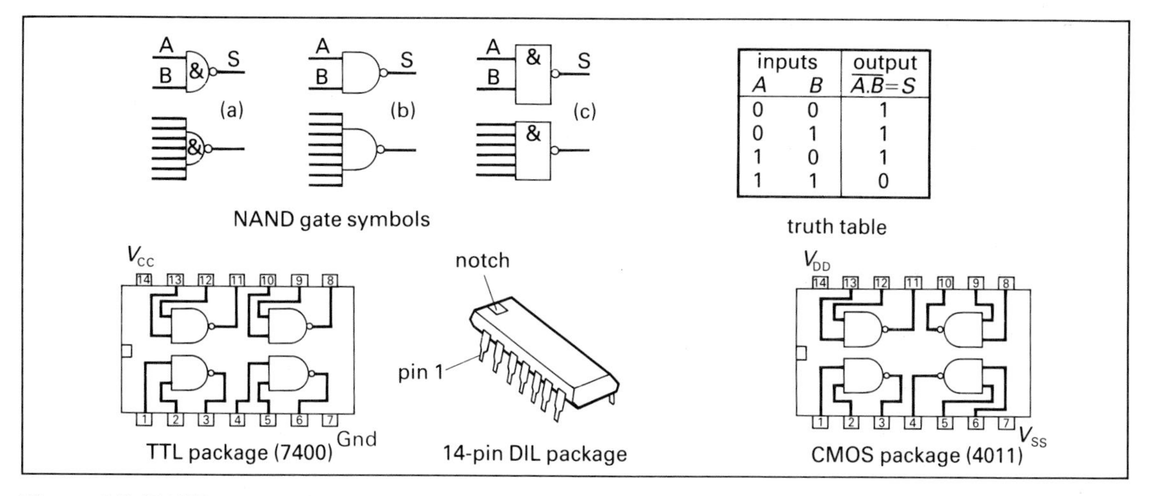

Figure 3.7 NAND gate symbols, truth table and packages

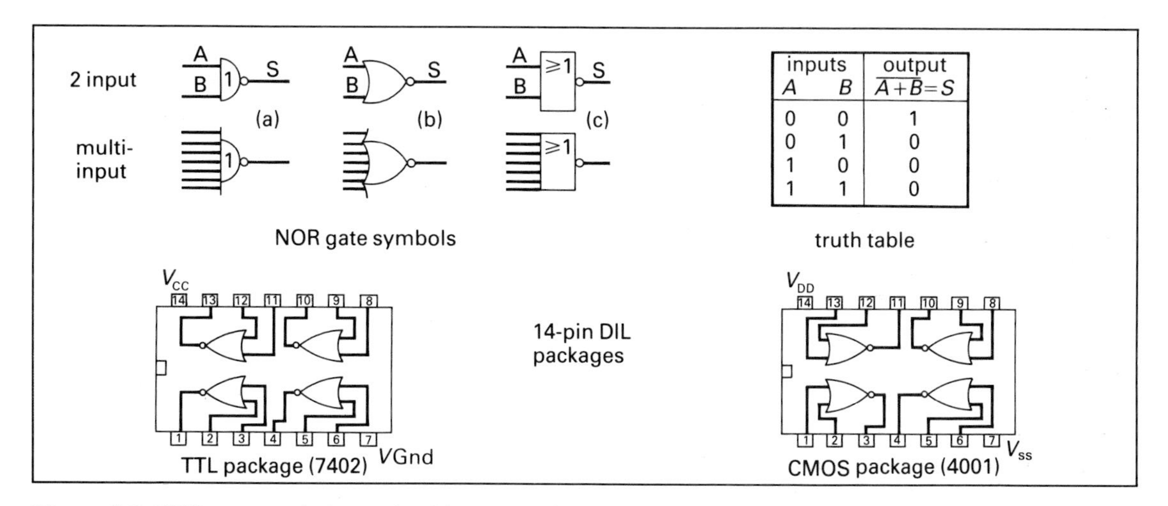

Figure 3.8 NOR gate symbols, truth table and packages

inputs are high. The output is high when both inputs are low. For multiple input NOR gates, a high on any input gives a low output; the output is high only when all inputs are low.

Figure 3.8 shows three types of circuit symbol used for NOR gates. These symbols are discussed in Section 3.12. The truth table gives the state of the output, 1 or 0, for all input combinations — hence the term combinational logic for logic systems designed using NOR gates.

Note the shorthand method of writing the function of a NOR gate. The statement 'not (input A or input B) gives output S' is written $\overline{A + B} = S$. The bar across the top of the OR expression, A + B, indicates that the output is obtained by inverting the output of the OR gate. This equation is a Boolean expression.

Examples of NOR gates in integrated circuit packages are shown in figure 3.8. These are quad 2-input NOR gates. Note that the CMOS gates are 'nose-to-nose' in the package, and the TTL gates are 'nose-to-tail'. Section 3.10 compares TTL and CMOS digital ICs.

Question

1 Write out the truth table for a 4-input NOR gate.

3.8 The exclusive-OR gate

This logic gate is sometimes referred to as the *any-but-not-all gate*. The term 'exclusive-OR' is sometimes shortened to XOR, pronounced 'exor'. The difference between the OR gate and the XOR gate is shown by the symbols and truth table in figure 3.9. Unlike the OR gate, the XOR gate gives an output of logic 0 when all inputs are at logic 1. Otherwise any input at logic 1 gives an output of logic 1. Thus for a 2-input XOR gate, the output is low when both inputs are high, but high if either input is high.

Figure 3.9 shows three types of circuit symbol used for XOR gates. These symbols are discussed in Section 3.12. The truth table gives the state of the output, 1 or 0, for all input combinations — hence the term combinational logic for logic systems designed using XOR gates.

Note the shorthand method of writing the function of a XOR gate. The statement 'input A or input B gives output S' is written $A \oplus B = S$. The symbol $\oplus$ means that the two inputs are 'XORed' together. This equation is a Boolean expression.

Examples of XOR gates in integrated circuit packages are shown in figure 3.9. These are quad 2-input XOR gates. Note that the CMOS gates are 'nose-to-nose' in the package, and the TTL gates are 'nose-to-tail'. Section 3.10 compares TTL and CMOS digital ICs.

Questions

1 Write down the truth table for an exclusive-NOR gate.

2 Is there anything wrong with the following advertisement for a job?
'Applicants should have two of the following qualifications: a driving licence; six GCSEs; experience of selling computers.'
What should the advert have said?

3 What single logic gate represents the following statement:
'It is an offence to drink and drive.'

4 How many horizontal rows are there in a truth table for a 4-input AND gate?

5 How can you get a 3-input AND gate using two 2-input AND gates?

6 A switch operates an electric motor which drives a machine. The switch is controlled by two light dependent resistors (LDRs) such that if either (or both) LDRs are covered, the motor stops running. Draw a logic diagram for this protection system.

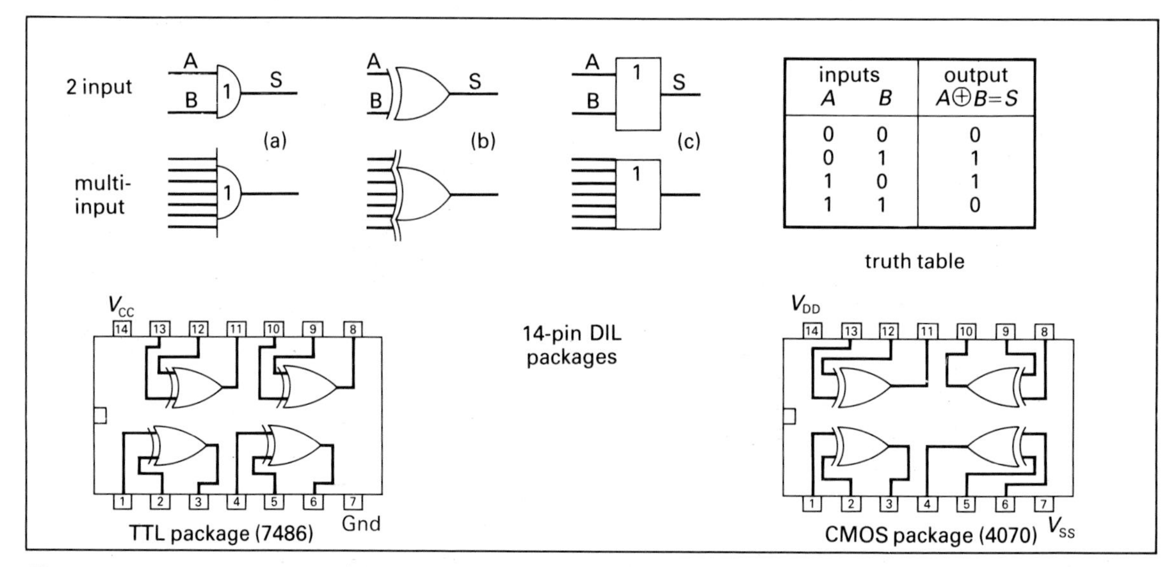

Figure 3.9 XOR gate symbols, truth table and packages

3·9 *Experiment* E2

Testing logic gates

Figure 3.10 shows how you can breadboard two simple circuits for testing the function of logic gates. The gates chosen as examples are:

Figure 3.10(a): an AND gate of the TTL family — a 7408 device.
Figure 3.10(b): a NAND gate of the CMOS family — a 4011 device.

Compare these two circuits with the packages shown in figure 3.4 (TTL AND gate), and figure 3.7 (CMOS NAND gate). In each case the two inputs to the gate are pins 1 and 2, and the output to each gate is pin 3.

Notice one important difference between the internal arrangements of the gates for the TTL and CMOS families: the internal arrangements of all quad CMOS gate packages follow a set pattern of pins: e.g. input pins 1 and 2, output pin 3; input pins 12 and 13, output pin 11 — a 'nose-to-nose' arrangement of gates in all the packages. So they can all (barring the 40106 inverter) be interchanged in a circuit leaving inputs and outputs to the gates unchanged. Though all quad TTL packages follow a 'nose-to-tail' arrangement, and the NAND, AND, and OR gates follow

(a)

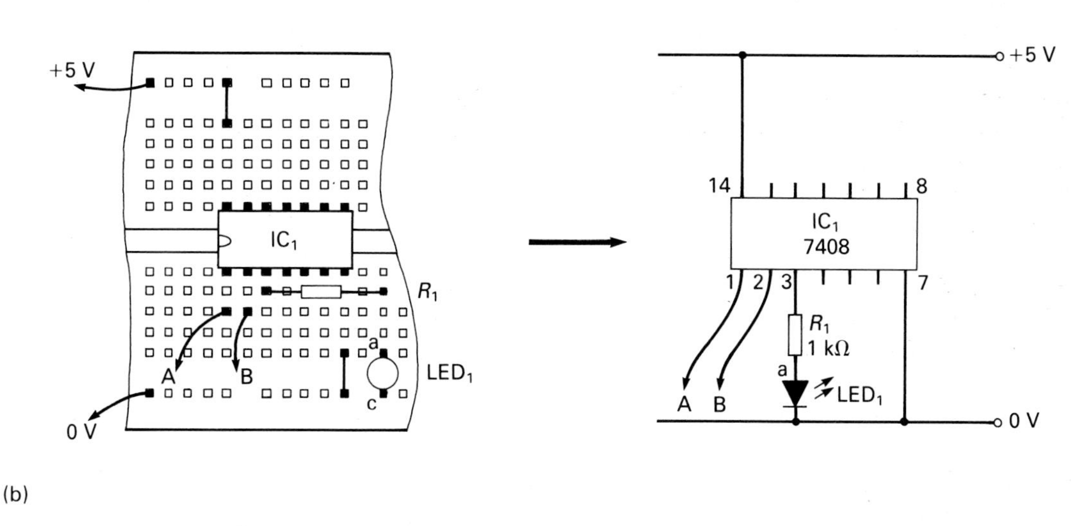

(b)

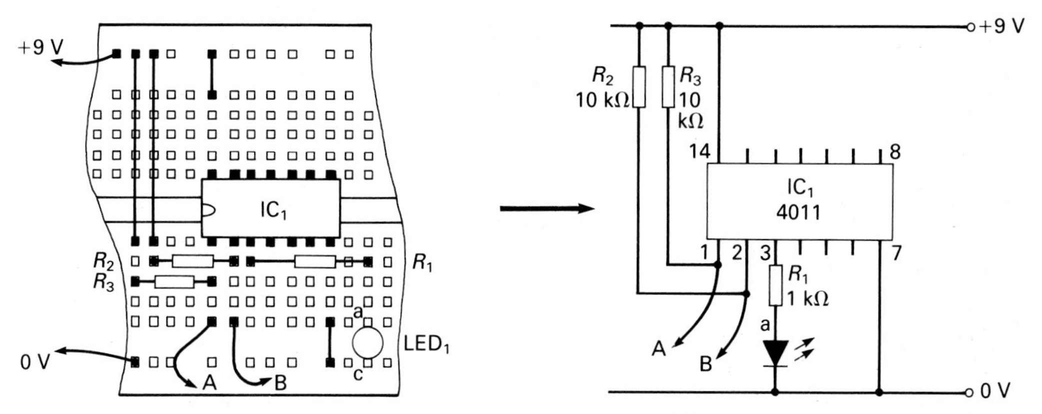

Figure 3.10 Experiment E2: Testing TTL and CMOS gates (a) AND gate using the TTL 7408 package; (b) NAND gate using the CMOS 4011 package

the same pattern in the packages, the TTL NOR gate is 'odd'! The lesson is: always check the internal arrangements of the gates for the logic devices you are using. Now we can go back to the experiment.

There are a number of rules about using the TTL and CMOS logic families which are compared in Section 3.10. Two of the rules are made clear in figure 3.10:

(a) TTL is operated from a nominal 5 V (+ 0.25 V) supply. CMOS operates from 3 V to 15 V, so 9 V is convenient for the latter.

(b) Unconnected TTL inputs are normally 'pulled' high by their internal circuitry. But unconnected CMOS inputs 'float' and need 'pull-up' or 'pull-down' resistors on their inputs to make sure the inputs are high (logic 1) or low (logic 0), respectively. Resistors R_2 and R_3 in figure 3.10(b) pull the two NAND gate inputs high; no such resistors are needed for the TTL gates.

LED_1 in each circuit checks whether the output of the gate is high (LED_1 on, i.e. logic 1), or low (LED_1 off, i.e. logic 0). The series resistor R_1 should not be less than 1 kΩ for CMOS gates but it can be as low as 270 Ω for TTL gates. The maximum output current for CMOS devices is much lower (about 1 mA) than for TTL devices (about 10 mA).

Since the TTL 7408 device is a quad 2-input AND gate, LED_1 should be lit (logic 1) if link wires A and B are left unconnected due to the effect of the internal circuitry. Connect link wire A only, then B only, then both A and B together, to 0 V. For each combination of inputs check the state of the output by referring to the truth table for the AND gate in figure 3.4.

Since the CMOS 4011 device is a quad 2-input NAND gate, LED_1 should be off (logic 0) if link wires A and B are left unconnected (logic 1 on each gate due to the effect of the 'pull-up' resistors, R_2 and R_3). Connect link wire A only, then B only, then A and B together, to 0 V. For each combination of inputs, check the state of the output by referring to the truth table for the NAND gate in figure 3.7.

3.10 Logic families

A 'family' is a particular design of logic integrated circuits such that all devices in the family will interact together and will drive each other's inputs satisfactorily. Today we use two main groups of logic family, the ones shown in figures 3.4 to 3.10: the TTL (transistor-transistor logic) '7400' series, and the CMOS (complementary metal-oxide semiconductor logic) '4000' series. The former is based on bipolar transistors and the latter on MOSFETs (Section 3.11). Figure 3.11 compares the main characteristics of the two families. There

Figure 3.11

characteristic	CMOS	TTL
normal supply voltage	3–18 V	+ 5 V
typical logic 1 output level	above 70% of supply	above 2.4 V
typical logic 0 output level	below 30% of supply	below 0.4 V
fan-out, i.e. the number of inputs a gate can drive	50 plus	10
typical power consumption per gate	less than 1 μW	22 μW
maximum usable frequency	2 to 4 MHz	50 MHz
handling	handle with care	no special precautions
input impedance	very high	fairly low
worst case noise immunity	30 to 40% of supply	1 V

are variations of both series, and the comparison refers to the two standard families.

The much lower power consumption, lower cost (especially for complex integrated circuits), wider range of operating voltage, and resistance to the effects of electrical noise, make CMOS devices a popular choice for designing digital systems. It's for these reasons that CMOS devices are chosen for the series of Project Modules in *Basic Electronics*.

The relatively slower speed of operation of CMOS compared with TTL is of no importance for the majority of applications with which we are concerned in *Basic Electronics*. Even so, if high operating speed is required, special high-speed CMOS, e.g. the 74HC series, are obtainable from some distributors. The lower output current of CMOS compared with TTL is a problem when driving capacitive loads. However, the use of the buffered B series, e.g. 4011B, helps solve this problem and gives a sharper and more powerful output voltage change than the standard 4000 series. If discrete transistors, e.g. VMOS transistors (see Book C, Chapter 26), are added to their outputs, CMOS can drive high-powered loads if required. If CMOS devices are handled with the care outlined in Book A, Section 12.2, they will not be damaged by static voltages. However, some protection against damage by static electricity is provided by two diodes and a series resistor in their input circuits. Incidentally, the range of CMOS devices includes analogue devices such as operational amplifiers (see Book D), as well as analogue switches, described in Section 8.5.

There are improved variants of the standard 7400 series of TTL devices. These include the progressively faster Schottky (the 74S series), low-power Schottky (the 74LS series), and the advanced low-power Schottky (the 74ALS series). There is also a high-speed CMOS 74HC series available in the standard 74 series pin-outs, similar in speed to the LS series but with 5 to 7 times smaller power consumption and operating on a supply voltage of 2 V to 6 V.

▽ 3.11 The structure of TTL and CMOS NAND gates

Many of the characteristics of the standard TTL and CMOS series of digital ICs can be gleaned by looking at the operation of the 2-input NAND gates shown in figure 3.12. TTL is based on bipolar transistors and CMOS on MOSFETs, i.e. field-effect transistors. These transistors are discussed in Book C, Chapter 26, and the operation of the standard CMOS NOT gate in Book C, Section 26.9.

(a)

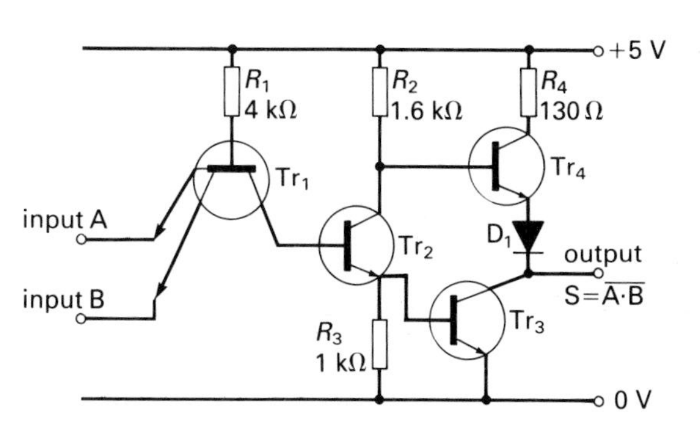

(b)

Figure 3.12 The structure of TTL and CMOS NAND gates: (a) TTL NAND gate (e.g. 7400 type); (b) CMOS NAND gate (e.g. 4011 type)

A particular characteristic of the TTL logic building block is the multi-emitter bipolar transistor, Tr_1, which performs the function of the two inputs, A and B, of TTL gates. Assume that these two inputs are logic high. Since the base of Tr_1 is held high via R_1, all the emitter-base junctions are reverse-biased. However, the base-collector junction of Tr_1 is forward-biased and current flows via R_1 through the base-emitter junctions of Tr_2 and Tr_3 turning them both on. So the output voltage is at logic low. Under these conditions, the voltages at the base of Tr_1 is about 2 V. If the input voltages are all higher than about 2 V, the output remains at logic low. In this state, Tr_3 acts as a *current sink* and will conduct up to 16 mA, i.e. ten standard TTL gates can be driven by the output of the TTL NAND gate (a fan-out of ten).

If input A or input B (or both inputs) go below 2 V, that particular emitterbase junction becomes forward-biased, and the base voltage of Tr_1 will fall to 0.9 V. This voltage is insufficient to forward-bias Tr_2 and Tr_3, which then turn off. The collector voltage of Tr_2 rises towards + 5 V and Tr_4 turns on, taking the output voltage via D_1 to + 3.3 V which is logic high at the output. R_4 limits the amount of current which flows if the output is accidentally short-circuited to 0 V. Thus, as the NAND gate truth table in figure 3.7 shows, the output is low if both inputs are high; if either or both inputs are low, the output is high. A 7400 TTL package contains four of these NAND gate circuits; a total of 16 transistors, 16 resistors and one diode all on a 2 mm square chip of silicon. This is an example of *small-scale integration* (SSI).

The CMOS NAND gate shown in figure 3.12 is made of two p-channel MOSFETs (Tr_1 and Tr_2) and two n-channel MOSFETs (Tr_3 and Tr_4). It is a variation of the CMOS NOT gate described in Book C, Section 26.9. The operation of the NAND gate can be explained with the help of the truth table in figure 3.13.

Figure 3.13

inputs		**state of MOSFETs**				**output**
A	**B**	**Tr_1**	**Tr_2**	**Tr_3**	**Tr_4**	**S**
0	0	on	on	off	off	1
0	1	on	off	off	on	1
1	0	off	on	on	off	1
1	1	off	off	on	on	0

A low output voltage can only occur if both n-channel enhancement MOSFETs, Tr_3 and Tr_4, conduct. This condition occurs only when both inputs A and B are high. If either or both inputs are low, either or both of the p-channel MOSFETs, Tr_1 and Tr_2, are off and the output is high. △

3.12 Logic symbols

The three types of logic symbols for the gates shown in figure 3.4 to 3.9 are all in use. The (a) symbols are 'British Conventional', the (b) symbols 'American Mil Spec', and the (c) symbols 'British Standard'. The American symbols are widely preferred since their different shapes are clearly recognised in complex circuit diagrams. However, there is some pressure in the United Kingdom to adopt the rectangular British Standard symbols. The American symbols are used in *Basic Electronics*.

The ampersand, '&', inside the AND and NAND boxes indicates the AND function; the small circle at the output of the NAND gate that the AND function is inverted, i.e. NOT-AND; the ≧ inside the OR gate indicates that the output is logic 1 for the case of both inputs at logic 1; the NOR gate shows the OR function is inverted by the small circle at its output; and the '= 1' inside the exclusive-OR gate, that the output is zero when both inputs are at logic 1.

4 Simple Boolean Algebra

4.1 Introduction

In 1847, George Boole invented a shorthand method of writing down combinations of logical statements which are either true or false. Boole proved that the binary or two-valued nature of logic is valid for symbols and letters instead of words.

Therefore Boolean algebra is an ideal method of analysing and predicting the behaviour of digital circuits which deal with on/off signals, since 'true' can be regarded as logic 1 (i.e. an 'on' signal), and 'false' as logic 0 (i.e. an 'off' signal). However Boolean algebra did not have an impact on digital electronics until 1938 when Shannon applied the new algebra to telephone switching circuits.

4.2 Boolean expressions for logic gates

Let's begin our look at Boolean algebra in action by summarising the Boolean expressions introduced in figures 3.4 to 3.9 for the seven basic logic gates shown in figure 4.1.

Remember

(i) Two symbols are ANDed if there is a dot between them.
(ii) Two symbols are ORed if there is a plus sign between them.
(iii) A bar across the top of a symbol means the value of the symbol is inverted.

Figure 4.1

logic statement	Boolean expression
(a) AND gate (figure 3.4) input A *and* input B = output S	$A.B = S$
(b) OR gate (figure 3.5) input A *or* input B (including input A *and* input B) = output S	$A + B = S$
(c) NOT gate (figure 3.6) *not* input A = output S	$\overline{A} = S$
(d) NAND gate (figure 3.7) *not* (input A *and* input B) = output S	$\overline{A.B} = S$
(e) NOR gate (figure 3.8) *not* (input A *or* input B) = output S	$\overline{A + B} = S$
(f) Exclusive-OR gate (figure 3.9) input A *or* input B (excluding input A *and* input B) = output S	$A \oplus B = S$
(g) Exclusive-NOR gate *not* (input A *or* input B, excluding *not* input A *and* input B) = output S	$\overline{A \oplus B} = S$

The bar has an important role to play in the shorthand of Boolean algebra. Its use is shown by the NOT gate (figure 3.6), i.e. $\overline{1} = 0$ and $\overline{0} = 1$. If the bar is applied twice, e.g. $\overline{\overline{1}}$, this means a double inversion. Thus $\overline{\overline{1}} = 1$ and $\overline{\overline{0}} = 0$.

Question

1 What is the value of $\overline{\overline{\overline{1}}}$?

4·3 Universal logic gates

Integrated circuit packages of NAND gates are not only cheaper to make than other gates, but one or more of them can be connected together to produce the logic functions of NOT, AND, OR, NOR, and XOR. That's why NAND gates are called universal gates; the same is true of the more expensive NOR gates. The rules of Boolean algebra can be used to show that the combinations of NAND gates really do give the other logic functions.

NOT gate (figure 4.2(a))
Since the two inputs to the NAND gate are connected together, we can write:

$\overline{A.B} = S$. If $A = B = 1$ then $\overline{1.1} = \overline{1} = 0$.

If $A = B = 0$ then $\overline{0.0} = \overline{0} = 1$.

So a NOT gate can be obtained by connecting together the two inputs to a 2-input NAND gate. A NOT gate is also obtained if all the inputs of a multiple input NAND gate are connected together.

AND gate (figure 4.2(b))
If a NAND gate is followed by a NOT gate, an AND gate is obtained (as you might expect!). Thus $R = \overline{A.B}$, and $S = \overline{R}$, so $\overline{\overline{A.B}} = S$, or $A.B = S$, since a double inversion leaves the expression unchanged.

OR gate (figure 4.2(c))
NAND gates 1 and 2 are each connected as NOT gates. Thus inputs Q and R to NAND gate 3 are: $Q = \overline{A}$ and $R = \overline{B}$. Thus $\overline{\overline{A}.\overline{B}} = S$. The truth table for this equation is shown in figure 4.3. Also shown is the truth table for an OR gate.

Figure 4.3 The truth table for an OR gate made from NAND gates

NAND gate combination						**OR gate**		
$\overline{\overline{A}.\overline{B}} = S$						$A + B = S$		
A	$\overline{A}$	B	$\overline{B}$	$\overline{A}.\overline{B}$	$\overline{\overline{A}.\overline{B}}$	A	B	A + B
0	1	0	1	1	0	0	0	0
0	1	1	0	0	1	1	0	1
1	0	0	1	0	1	0	1	1
1	1	0	0	0	1	1	1	1

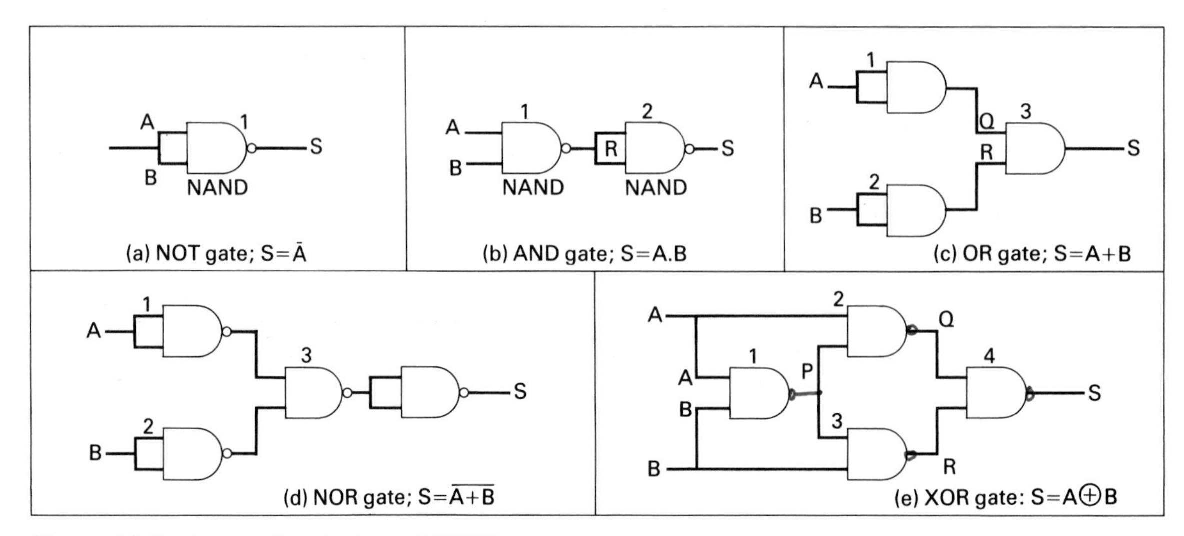

Figure 4.2 Logic gates from 'universal' NAND gates

The truth tables are identical. Thus we have to conclude that the combination of NAND gates shown in figure 4.2(c) is equivalent to an OR gate. Moreover, the Boolean expressions must be equivalent to each other, i.e. $\overline{\overline{A}.\overline{B}} = A + B$, and a very useful expression this is for analysing the function of digital circuits. It is one of two statements known as De Morgan's Theorems which are useful in analysing logic circuits as explained in Section 12.1.

NOR gate (figure 4.2(d))
This one is easy since the OR gate is followed by an inverter to give the NOR function. Thus we add a seventh column to the NAND gate truth table, and a fourth column to the OR gate truth table above as follows.

Figure 4.4 The truth table for a NOR gate made from NAND gates

$\overline{A}.\overline{B}$	$\overline{A + B}$
1	1
0	0
0	0
0	0

Since these two columns are identical, we have to conclude that the combination of NAND gates in figure 4.2(d) performs the function of a NOR gate. Moreover, the Boolean expressions are equivalent to each other, i.e. $\overline{A}.\overline{B} = \overline{A + B}$. This expression is one of two know as De Morgan's Theorems which are explained in Section 12.1.

Figure 4.5 The truth table for an XOR gate made from NAND gates

NAND gate combination								XOR gate
A	B	$P = \overline{A.B}$	$A.\overline{A.B}$	$B.\overline{A.B}$	$Q = \overline{A.\overline{A.B}}$	$R = \overline{B.\overline{A.B}}$	$S = \overline{(\overline{A.\overline{A.B}}).(\overline{B.\overline{A.B}})}$	$S = A \oplus B$
0	0	1	0	0	1	1	0	0
0	1	1	0	1	1	0	1	1
1	0	1	1	0	0	1	1	1
1	1	0	0	0	1	1	0	0

XOR gate (figure 4.2(e))
It is much more difficult to prove that this arrangement of NAND gates is equivalent to an XOR gate, but work through the truth table in figure 4.5 and you will find it is all very logical!

Questions

1 What are the meanings of the logic diagrams in figure 4.6? Write down the Boolean expressions for each logic diagram.

(a)

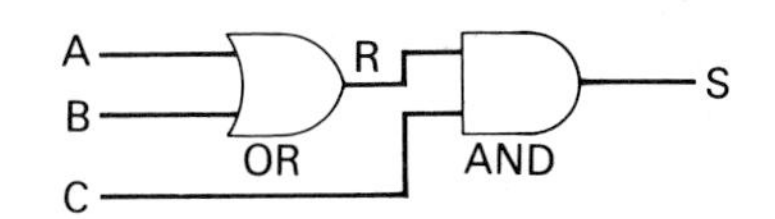

(b)

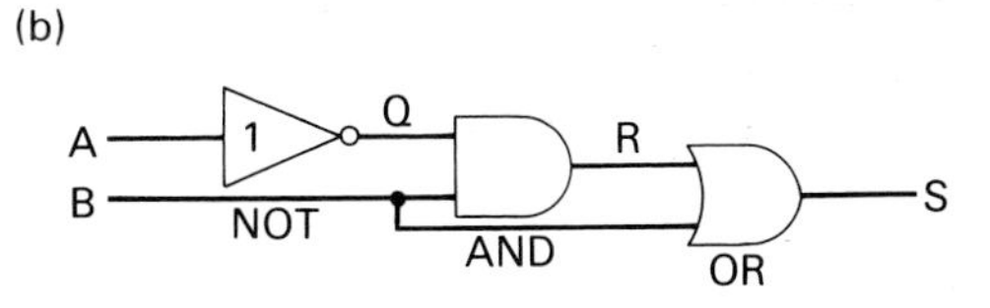

Figure 4.6 Two logic functions

2 Write down the truth tables which give the logic state of the output, S, for the above logic diagrams.

3 Write down the truth table for the logic diagrams in figure 4.7.

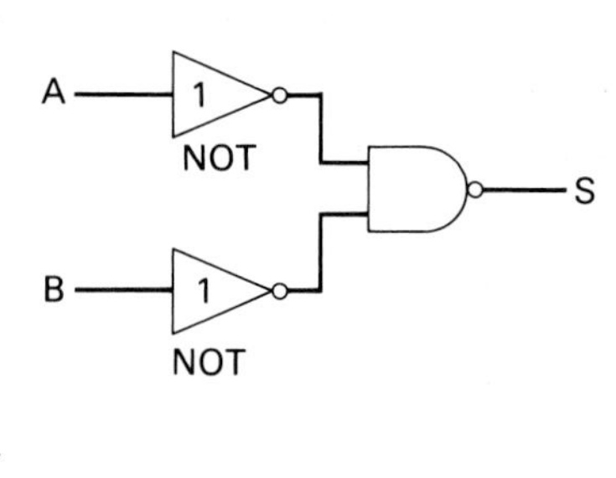

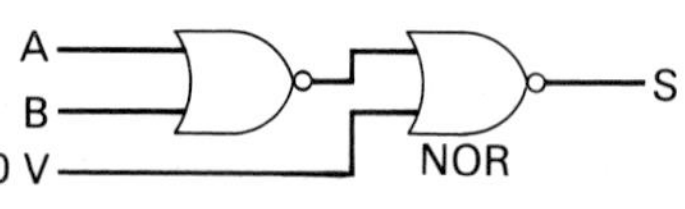

Figure 4.7 Two more logic diagrams

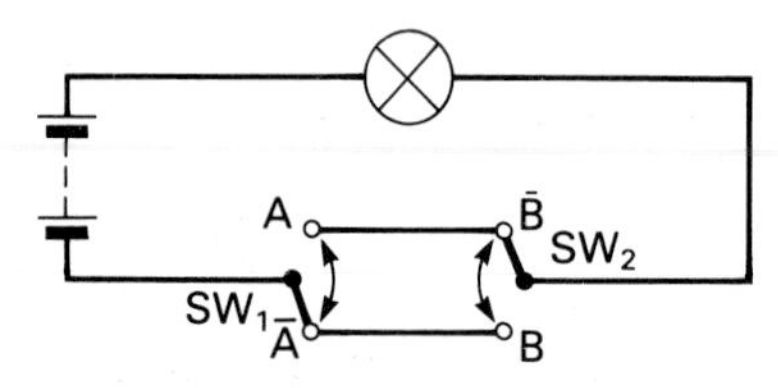

Figure 4.8 Two-way light circuit

4 A logic diagram has inputs A, B and C. Write down its truth table if its output, S, is given by S = A + B.C.

5 Figure 4.8 shows a two-way lighting circuit. The two positions for each switch are given logic values of A and $\overline{A}$, B and $\overline{B}$. Draw up a truth table for this circuit, using lamp on — logic 1; lamp off — logic 0. What type of logic function is this circuit?

4.4 *Experiment* E3

Designing logic gates from NAND gates

The 7400 TTL package, and the CMOSL 4011 package each contain four NAND gates. Figure 4.9(a) shows how two of the 7400 NAND gates can be used to produce an AND gate — the same circuit as figure 4.2(b). Figure 4.9(b) shows how four of the 4011 NAND gates can be used to produce an OR gate — the same circuit as figure 4.2(c).

Check the truth tables you obtain with these two circuits with those in figures 3.4 and 3.5. Also make an exclusive-OR gate from all four NAND gates. How would you make an exclusive-NOR gate with NAND gates?

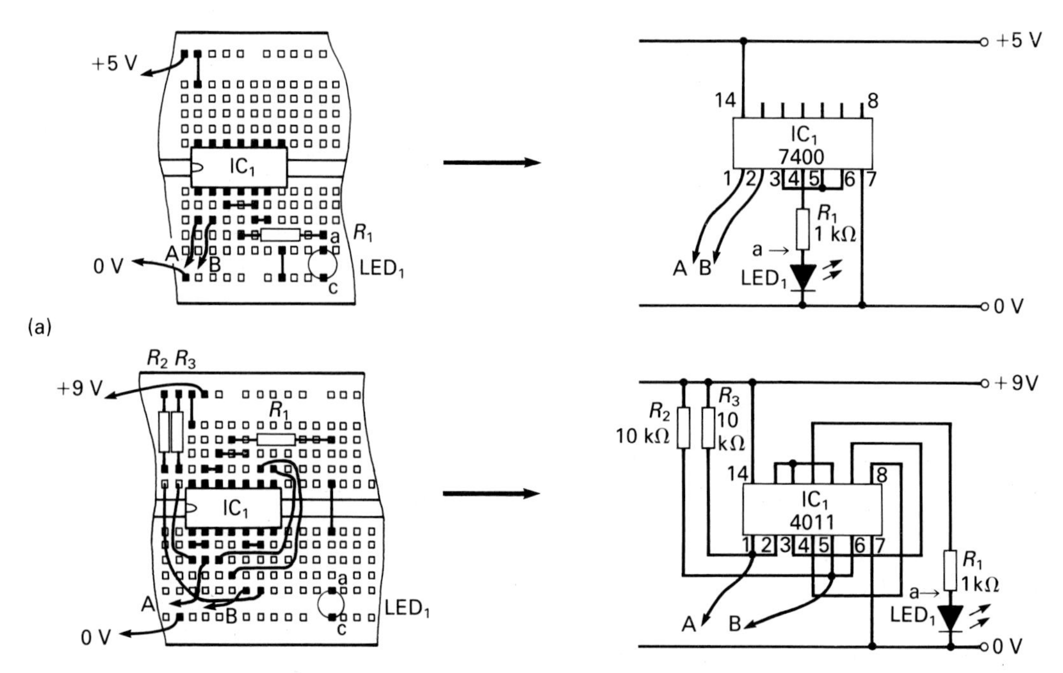

Figure 4.9 Experiment E3: (a) TTL AND gate from 7400 NAND gates; (b) CMOS OR gate from 4011 NAND gates

5 Flip-flops and Binary Counters

5.1 Combinational and sequential logic

As Chapter 3 shows, the truth table for a logic gate tells us exactly what happens at the gate's output, i.e. logic 0 or 1, given a particular combination of digital signals at its inputs. That's why a decision-making circuit using gates is generally called *combinational logic*. The use of combinational logic in encoders, decoders and multiplexers is described in Chapters 7 and 8. Chapter 14 shows how they are used in arithmetic circuits.

Flip-flops are made from logic gates, and their function in digital circuits is two-fold: they are used in *binary counters*, *dividers* and *shift registers*, and they are used to store binary data, i.e. to provide *electronic memory*. A flip-flop remembers its binary data until it is 'told' (electronically) to forget it. The name *sequential logic* generally describes the operation of digital circuits based on flip-flops. A sequential logic circuit is one that has memory, and therefore gives an output that depends on previously stored bits, i.e. 0s and 1s, of information as well as on any new data.

In the sections which follow, we look at the way TTL and CMOS flip-flops are used in binary counters, dividers and shift registers. Section 15.3 shows how they are used in the random-access memory (RAM) of computer systems. Note that the flip-flop is also called a *bistable*, since it is one of the family of multivibrators (see Chapter 9), and a *latch* since it 'catches hold' of data until it is instructed to change it. The Bistable (Project Module B3), the Decade Counter (Project Module C3), and the Frequency Divider (Project Module D5) are examples of integrated circuits based on flip-flops.

5.2 Flip-flop symbol

The circuit symbol used for a JK flip-flop is shown in figure 5.1. The logic pulses which operate the flip-flop enter at the clock (CLK) terminal where the small circle indicates that the flip-flop works when the voltage of the input clock pulse changes from high to low. The significance of the other two inputs, J and K, is explained in Section 13.4 so don't worry about their purpose now.

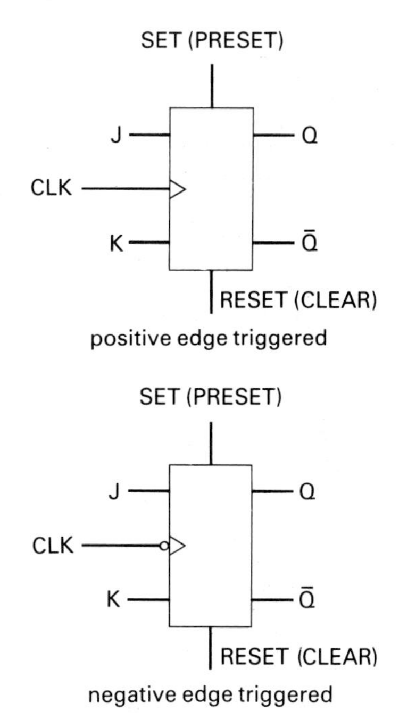

Figure 5.1 Symbols of flip-flops

The two connections SET and RESET, or PRESET and CLEAR, are used to set the outputs, Q and $\overline{Q}$, to a desired state. Thus if a logic 1 or 0 (depending on the type of flip-flop used), is put on the SET (or PRESET) terminal the output states of the flip-flop change so that output Q goes to low and $\overline{Q}$ to high.

The bar over the $\overline{Q}$ output means that the output is *not* the Q output, i.e. if $Q = 0$, $\overline{Q} = 1$, and vice versa. The Q and $\overline{Q}$ outputs are called *complementary outputs*. Our main interest at present is in the operation of the flip-flop when the Q and $\overline{Q}$ outputs are *toggled* by the clock for this is the basis of binary counters and dividers. 'Toggling' means making the complementary outputs, Q and $\overline{Q}$, change their logic states for each high to low (or low to high) change of the clock pulse. The effect on the high/low states of these outputs as the clock pulse regularly goes high, low, high, etc. is shown in figures 5.2 and 5.3.

5·3 *Experiment* E4

Toggling a flip-flop

Figures 5.2 and 5.3 show the 14-pin TTL 7476, and the 16-pin 4027 devices. These packages contain two JK flip-flops. Also shown are the TTL 7414 and the CMOS 40106 packages which are used to obtain the clock pulses to toggle the flip-flops. These devices contain six (hence 'hex') Schmitt triggers.

Use either the TTL (figure 5.2) or the CMOS (figure 5.3) circuits. A breadboard assembly of the CMOS circuit is shown. Clock pulses at a frequency of about 1 Hz (seen by means of LED_1) are obtained by wiring up one of the Schmitt triggers, IC_1, as an oscillator. These pulses are fed into the flip-flop via its CLK terminal. LED_2 and LED_3 indicate what happens to the Q and $\overline{Q}$ outputs. Note the following changes

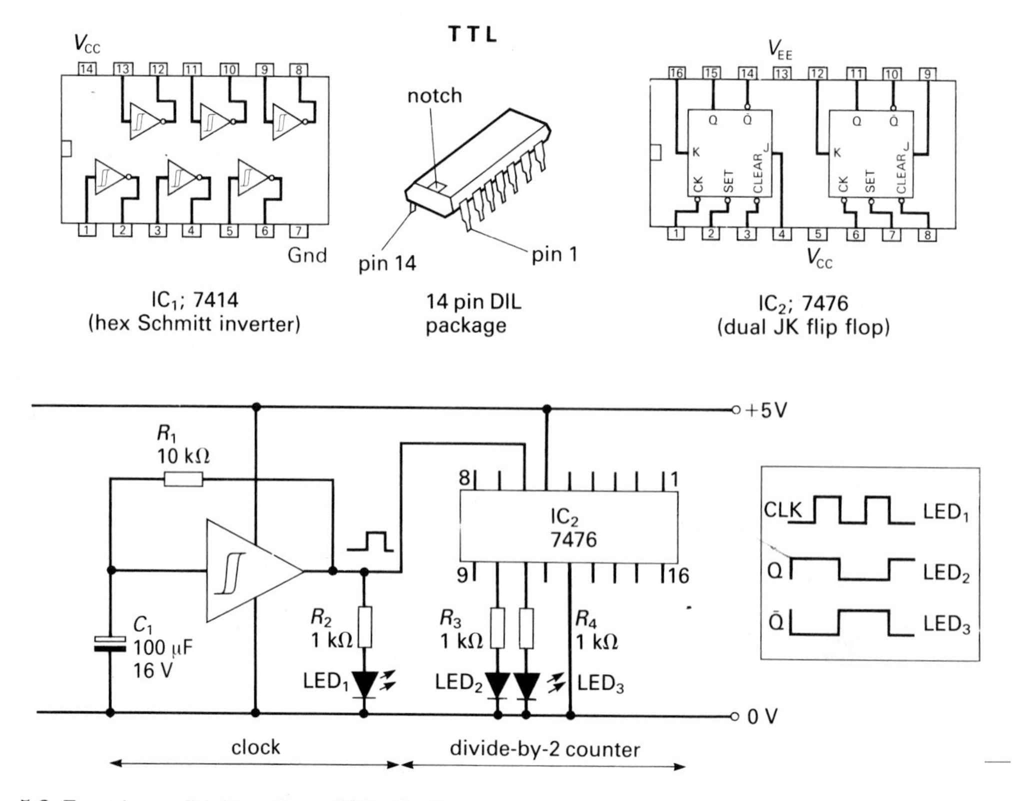

Figure 5.2 Experiment E4: Toggling a TTL flip-flop

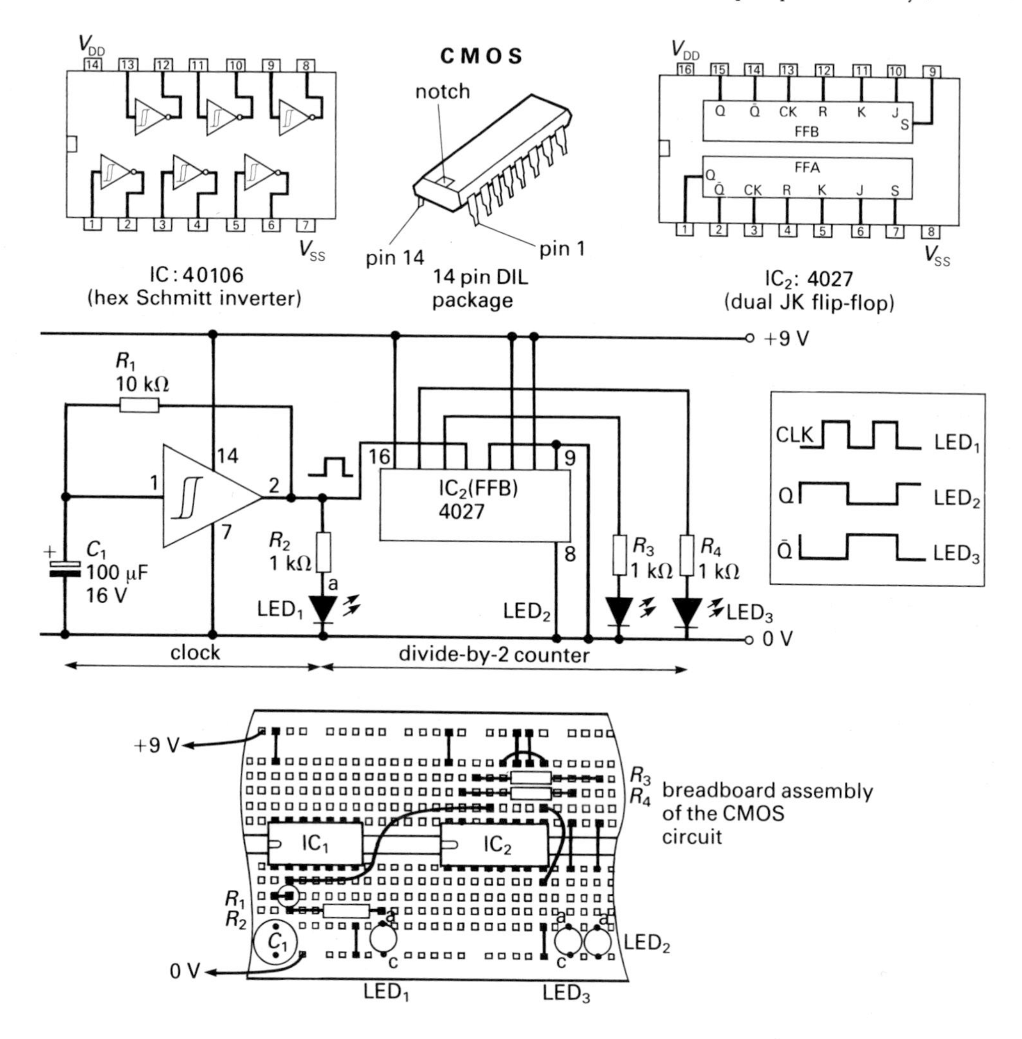

Figure 5.3 Experiment E4: Toggling a CMOS flip-flop

in the Q and $\overline{Q}$ outputs, where H represents a high signal and L a low signal.

Figure 5.4

CLK	Q	$\overline{Q}$
L(0)	H(1)	H(0)
H(1)	H(1)	L(0)
L(0)	L(0)	H(1)
H(1)	L(0)	H(1)
L(0)	H(1)	L(0)
H(1)	H(1)	L(0)

. . . . and so on

(a) When LED_1 goes off (high to low change of clock pulse), LED_2 comes on (Q high) and LED_3 goes off ($\overline{Q}$ low). The Q and $\overline{Q}$ outputs are said to toggle, going on and off on successive high to low changes of the clock pulse.

(b) LED_2 and LED_3 stay on and off for twice the length of the clock pulse. Thus each output of the flip-flop divides the frequency of the clock pulse by two. The truth table in figure 5.4 shows the high (1) and low (0) states of the Q and $\overline{Q}$ outputs as the flip-flop is toggled by the clock.

(c) Note that the flip-flop stores a bit of data, e.g. 1, until the input clock pulse changes the data to 0. Thus it acts as a 1-bit memory, holding data until told to change it. Remember that the flip-flop is also called a bistable because it has two stable states, Q and $\overline{Q}$ which take logic values of 1 or 0.

(d) A single flip-flop acts as a divide-by-two counter when it is made to toggle. Thus if the input frequency is 1 kHz, the frequency at the Q and $\overline{Q}$ outputs is 500 Hz. How could we divide the frequency by two again?

5.4 *Experiment* E5

Building a 2-bit binary counter

Since the 7476 and the 40106 packages contain two flip-flops, FF_A and FF_B, what happens if the Q output of FF_A is fed into the CLK input of FF_B? Figure 5.5 shows how this is done using the 4027 CMOS flip-flop. In this circuit, LED_2 and LED_3 are used to monitor the Q outputs of FF_A and FF_B. Note that the Q output from FF_A (pin 3) is connected to the CLK input (pin 13) of FF_B. Thus the pulses at the Q output of FF_B have a quarter of the frequency of the clock pulses into FF_A, i.e. FF_B divides the clock frequency by two again. So if the clock frequency is 1 kHz, 500 Hz is obtained from the Q output of FF_A and 250 Hz from the Q output of FF_B.

Make sure that link wire, L, is connected to the 0 V rail on the breadboard. It holds the RESET pins, 4 and 12, of FF_A and FF_B at 0 V to allow the flip-flops to toggle. If you remove this wire from 0 V, resistor R_5 pulls pins 4 and 12 to + 9 V and resets the two-bit counter to zero, i.e. both LEDs go off. Replace the link wire and the counter begins to count from zero. All counters have this RESET facility.

Here we have a two-bit binary counter.

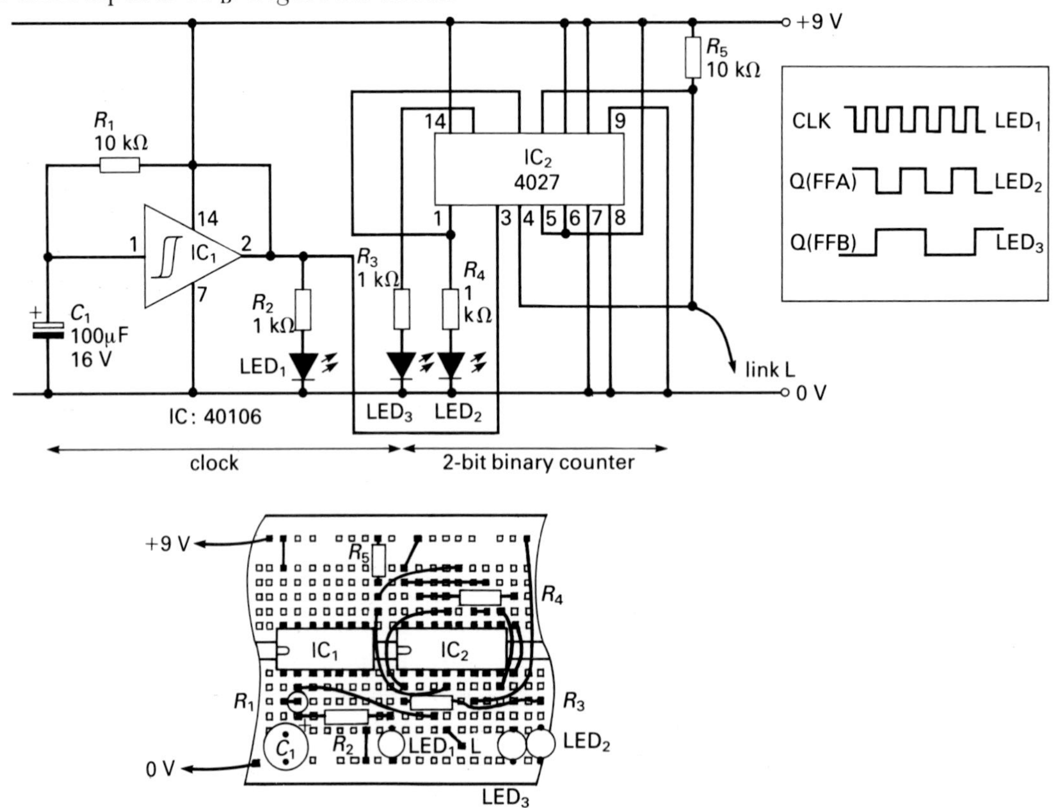

Figure 5.5 Experiment E5: Two-bit binary counter using two flip-flops

LED_2 and LED_3 go through the following sequence of highs (logic 1s) and lows (logic 0s).

Figure 5.6

CLK	Q (FF_A)	Q (FF_B)
H(1)	L(0)	L(0)
L(0)	H(1)	L(0)
H(1)	H(1)	L(0)
L(0)	L(0)	H(1)
H(1)	L(0)	H(1)
L(0)	H(1)	H(1)
H(1)	H(1)	H(1)
L(0)	L(0)	L(0)
H(1)	L(0)	L(0)
L(0)	H(1)	L(0)
H(1)	H(1)	L(0)
L(0)	L(0)	H(1)
8 clock states gives →	4 FF_A states gives →	2 FF_B states

LED_2 and LED_3 count four input pulses, 0 to decimal 4 in binary. The sequence is as follows (binary 1 = LED on; binary 0 = LED off):

Figure 5.7

LED_3 most significant bit (m.s.b.)	LED_2 least significant bit (l.s.b.)	input pulse (decimal count)
0	0	0
0	1	1
1	0	2
1	1	3

The next pulse sets the count to 00.

Clearly, to count beyond decimal 4 in binary we need further flip-flops. The CLK input to the next flip-flop, FFC, would be taken from the Q output of FF_B, and so on. Each flip-flop would add another bit to the binary number. Four flip-flops would produce a 4-bit binary counter which would count to decimal 15, 0000 to 1111. As flip-flops are cascaded in this way, repeated division by two occurs, making up the binary count. If there are *n* flip-flops, there are *n* bits in the number which can be counted.

As you can see from figure 5.5, the addition of individual flip-flops, or even packages of dual flip-flops, makes for a rather complicated breadboard circuit. Fortunately, four flip-flops can be obtained in a single IC package as explained below. (If you are unsure about counting in binary, Chapter 10 will help.)

5·5 *Experiment* E6

Building 4-bit and BCD counters

Figure 5.8 shows TTL and CMOS IC packages which contain four cascaded flip-flops so that they can count a 4-bit binary number. The TTL 7493 and the CMOS 4516 4-bit counters count up to decimal 15 (binary 1111). The 7490 and 4510 binary-coded decimal (BCD) counters count up to decimal 10 (binary 1010). BCD counters are particularly useful devices in circuits designed to display decimal numbers on seven-segment displays — see Chapter 7. Chapter 11 explains 4-bit binary codes.

The 7490 and 7493 devices are made more versatile by having one flip-flop separately connected so that they can be used independently as a scale-of-two counter while the remaining flip-flops can act as a scale-of-five counter. The CMOS devices can be used to count up or down in binary using the up/down (U/D) pin number 10. Note that the four outputs of these counters are designated Q_A, Q_B, Q_C and Q_D. Generally, Q_D is the most significant bit (m.s.b.) and Q_A is the least significant bit (l.s.b.) in the 4-bit count.

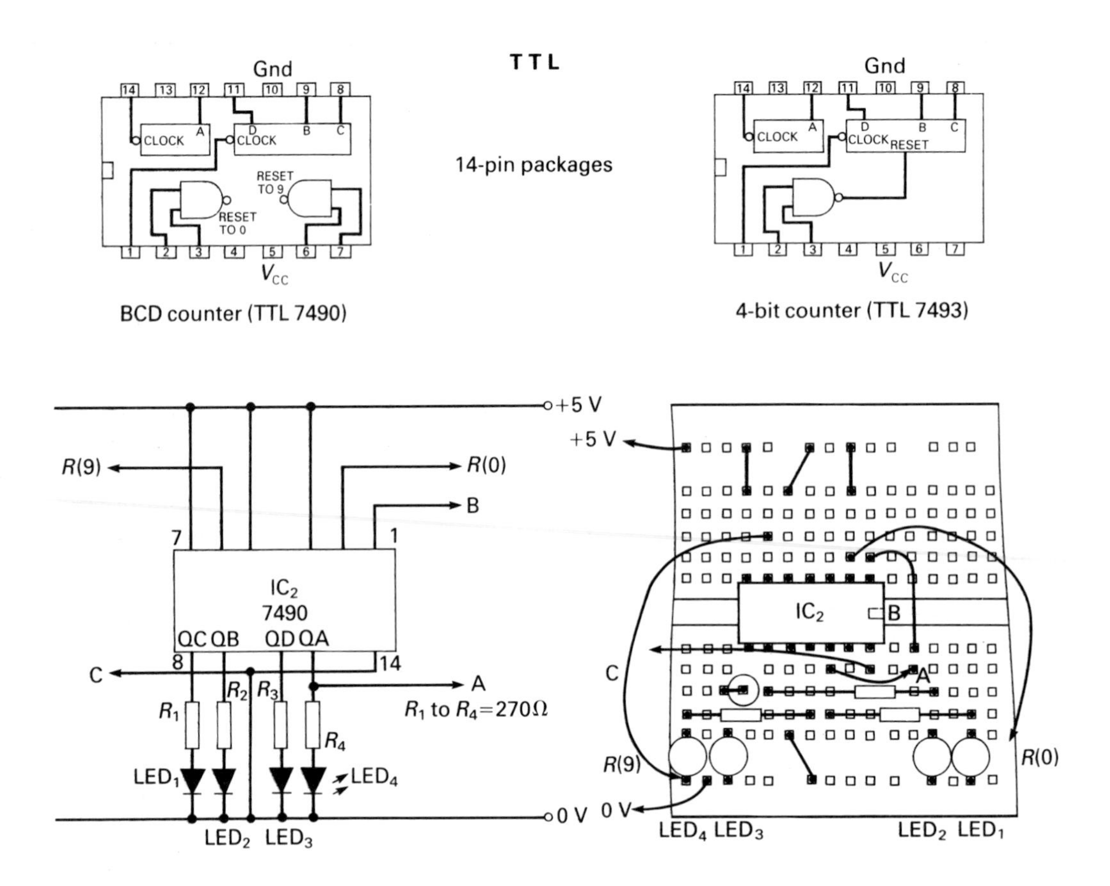

Figure 5.8 Experiment E6: Testing the 7490 BCD counter

Check the following ways of using the TTL 7490 and 7493 counters by assembling the circuit of figure 5.8 on breadboard. Regular pulses to 'clock' the counters are obtained from the Schmitt trigger oscillator of figure 5.3.

BCD count using the 7490
Link the output of flip-flop A (pin 12) to the input of flip-flop B (pin 14). Link R(0) and R(9) to 0 V. The 7490 now gives a BCD count, the outputs repeatedly passing through the cycle shown in figure 5.9.

When the LEDs display a count other than 0000, remove link R(0) from the 0 V rail. This will reset the counter to 0000. When the link is replaced, counting continues. The reset operates via the NAND gate inside the IC. One input (pin 3) of this NAND gate is held high by a

Figure 5.9

CLK pulse	Q_D LED_4 (m.s.b.)	Q_C LED_3	Q_B LED_2	Q_A LED_1 (l.s.b.)
0	0	0	0	0
1	0	0	0	1
2	0	0	1	0
3	0	0	1	1
4	0	1	0	0
5	0	1	0	1
6	0	1	1	0
7	0	1	1	1
8	1	0	0	0
9	1	0	0	1

direct connection to + 5 V. The other input (pin 2) automatically goes high when R(0) is removed, and this makes the output of the NAND gate go to logic 0.

Reconnect R(0) to 0 V. Now remove link R(9) from the 0 V rail and this will reset the counter to decimal 9 (binary 1001).

4-bit counter using the 7493

Disconnect the 5 V supply and replace the 7490 by the 7493 and clock it using the same circuit as for the 7490. This time, link R(9) does not work since the 7493 does not have a NAND gate inside it as does the 7490. However, link R(0) works to reset the counter to zero as for the 7490.

The 4510 and 4516 BCD and 4-bit counters

These two counters are more versatile than their TTL counterparts, and are housed in 16-pin packages which have identical pin arrangements as shown in figure 5.10. It is interesting to build these counters on breadboard because they can count up by holding pin 10 high or count down by holding pin 10 low. The following operating rules apply if you assemble these counters on breadboard.

(a) For normal BCD (4510) or 4-bit (4516) operation, connect CI (CARRY-IN), R (RESET), and PE (PRESET ENABLE) to + V, i.e. high. The count advances by binary 1 on each low to high change of the clock pulse if the U/D pin is high. Or the count decreases by binary 1 if the U/D pin is low. At any count if the R (RESET) pin is taken low, the counter resets to 0000.

(b) If CI is taken low, the counter stops counting. If the PE (PRESET) pin is taken high a binary number on the A, B, C and D inputs is loaded into the counter and appears at the Q0, Q1, Q2, and Q3 outputs.

(c) Note that the CI (CARRY-IN) pin is used to take the count from the CO (CARRY-OUT) pin of a preceding counter when the 4520 and 4516 counters are cascaded for counting beyond decimal 15.

▽ 5.6 The modulus of a counter

The modulus of a counter is the number of output states it goes through before resetting to zero. Thus a counter constructed from three flip-flops is a 'modulus 8' counter since it has eight output states representing the decimal numbers 0 (binary 000) to 7 (binary 111). A counter with *n* flip-flops is a modulus *n* counter.

Question

1 What is the modulus of the TTL 7493 counter?

CMOS 4510 counter

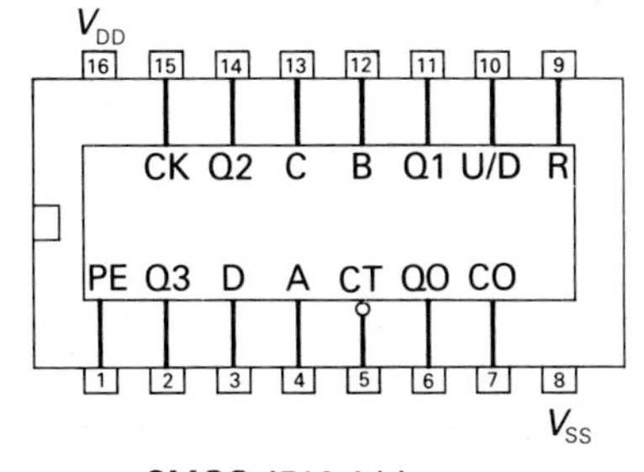

CMOS 4516 4-bit counter

Figure 5.10 Pin connections for the 4510 and the 4516 packages

Sometimes we need counters which have moduli other than 2, 4, 8, 16, etc. A smaller modulus counter can always be constructed from a larger modulus counter by skipping output states. Such counters are said to have a 'modified count'. How many flip-flops are required to produce a modified count? First find the next highest 'natural count', e.g. the number 4 is the next highest natural count of a mod 3 counter. Then, because two flip-flops are required to produce a natural count of 4, a minimum of two flip-flops is required to produce a mod 3 counter.

Question

2 How many flip-flops are required to produce
(a) a mod 6 counter;
△ (b) a mod 9 counter?

▽ 5·7 Cascaded BCD counters

If the 7490 and 4016 BCD counters are called upon to count beyond decimal 9 (binary 1001), two or more counters have to be cascaded as shown in figure 5.11. Initially all the counters are set to 0000 by applying a reset pulse to all counters simultaneously. The condition of all three counters is now

0000	0000	0000
hundreds	tens	units

After 9 clock pulses, the 'units' counter has an output state of 1001. On the tenth clock pulse, the Q_D output of the units counter changes from 1 to 0. This high to low transition triggers the 'tens' counter and the units counter returns to 0000. Thus

0000   0001   0000

As ten further clock pulses arrive, the units counter advances to 1001 again and on the next pulse the tens counter reads 0010. Thus after 99 clock pulses the BCD counters read

0000   1001   1001

After the next clock pulse, the Q_D output from the tens counter changes from high to low and the counter reads

0001   0000   0000

and so on. The maximum number of clock pulses which can be counted by these three cascaded BCD counters is 999. Cascaded BCD counters like those shown in figure 5.11 are commonly used in frequency counters, digital voltmeters, etc. The decoder/driver circuits introduced in Chapter 7 show how the accumulated count on each BCD counter can be displayed on a seven-segment display.

Questions

1 Five 7490 BCD counters are cascaded.
(a) What is the highest decimal count possible?
(b) If the 7490s are reset and 6425 pulses are received, what are the readings of the BCD outputs?

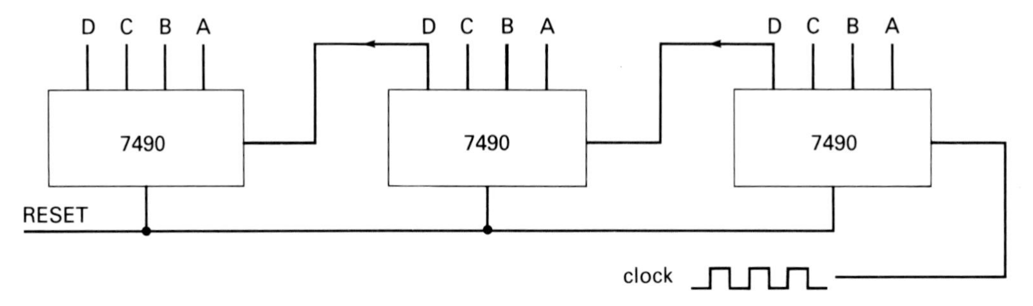

Figure 5.11 Cascaded BCD counters

2 An electronic wristwatch has a crystal clock which operates at 32 kHz.
(a) By what factor must this frequency be divided to obtain an output pulse of frequency 1 Hz?
(b) Given a number of flip-flops, 4-bit binary counters and BCD counters, draw a block diagram to show how you would obtain a 1 Hz pulse. △

▽ 5.8 Gating a counter

Gating a counter means turning it on for a specified period of time during which it counts the number of pulses that arrive at its input. Figure 5.12 shows a simple way of gating a counter. Here a 2-input AND gate drives a number of cascaded BCD counters. The train of pulses to be counted enters one of the inputs to the AND gate while a rectangular gating pulse is applied to the other input. The gating pulse resets the counter to zero at time t_1. Immediately after, when the gating pulse is high, the gating pulse enables the AND gate and allows clock pulses to pass to the counter. At time t_2 the AND gate is disabled and the counter stops counting, leaving the count displayed.

If the count displayed is to be accurate and stable, the gating pulse must be of precise and stable width. One way of providing gating pulses is to divide by ten repeatedly, a stable 5 MHz frequency. Using seven cascaded divide-by-10 BCD counters, the gating pulse would have a precise width of 2 s.

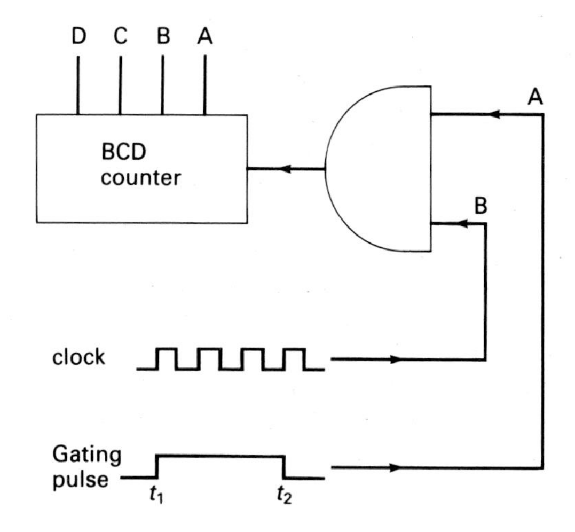

Figure 5.12 Gating a BCD counter

Question

1 Five cascaded BCD counters are set to zero, and a gating pulse exactly 2 s wide allows pulses at 32 kHz to enter the counters. What are the outputs of the five counters at the end of the gating pulse? △

6 Frequency Dividers

6.1 Dividing the mains supply frequency

A common use for counters is for frequency division. We have seen in Chapter 5 how the BCD counter can divide an input pulse by 10, or 7 or another required factor. An example of a simple divider as the basis for an electronic clock is shown in figure 6.1. The 50 Hz input frequency from the mains supply (changed from a sinusoidal to a square wave) is fed to a frequency divider which must divide by 50 to produce 1 Hz pulses. A divide-by-10 counter would produce a 5 Hz square wave. If this is followed by a divide-by-5 counter a 1 Hz output signal is obtained.

A practical circuit for obtaining 1 Hz pulses from the mains supply is shown in figure 6.2. Frequency division by 50 is provided by two TTL BCD 7490 counters, IC_3 and IC_4. IC_3 divides by 10, and IC_4 divides by 5. A stabilised 5 V supply for these counters is provided by the power supply based on the low-voltage mains transformer, T_1, and the voltage regulator, IC_1 (Book C, Section 9.3 discusses this circuit).

The Schmitt trigger IC converts the unrectified 50 Hz sinusoidal signal from the transformer to a 50 Hz square wave signal. This signal is fed to flip-flop B in the BCD counter, IC_3, which is wired to divide by 10. Compare the wiring of IC_3 with the pin connections for the 7490 BCD counter shown in figure 5.8. The 50 Hz signals are input (pin 1) to flip-flop B. The output (pin 11) of flip-flop D is one fifth of the input frequency, i.e. 10 Hz, and this is

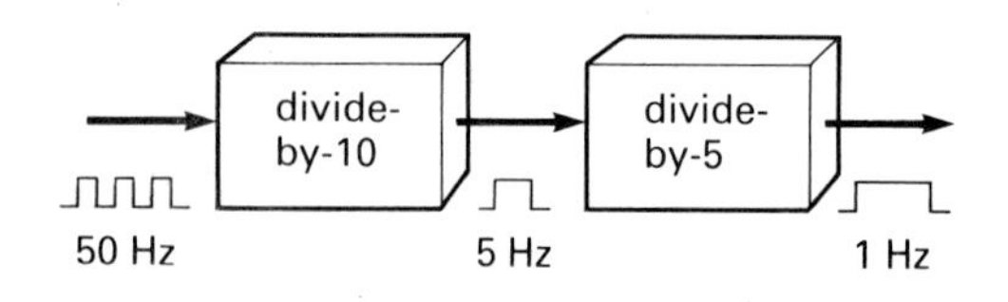

Figure 6.1 Using counters to obtain 1 Hz from 50 Hz main supply frequency

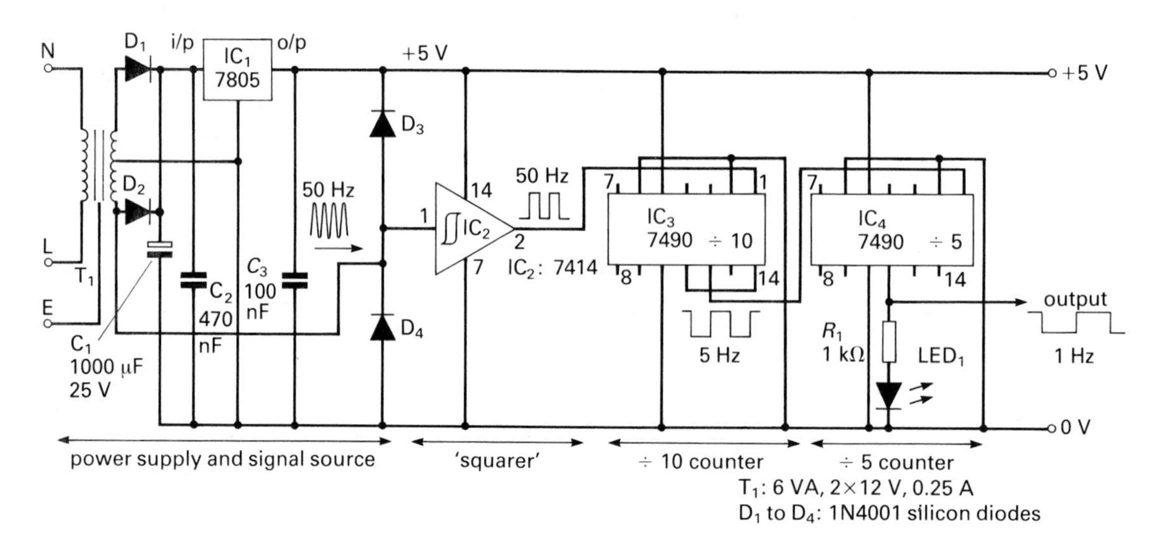

Figure 6.2 Practical circuit for 1 Hz pulses

fed to the input (pin 14) of flip-flop A to provide a further division by 2 to give an output (pin 12) of 5 Hz, one tenth of the input frequency. The following truth table shows the ten logic states of the outputs of IC_3 and is known as a *bi-quinary count.*

Figure 6.3

count input to pin 1	Q_D	Q_C	Q_B	Q_A
0	0	0	0	0
1	0	0	1	0
2	0	1	0	0
3	0	1	1	0
4	1	0	0	0
5	0	0	0	1
6	0	0	1	1
7	0	1	0	1
8	0	1	1	1
9	1	0	0	1
10 input counts				1 output count

A further division by 5 is obtained with the second 7490, IC_4, which is wired as a mod 5 counter, i.e. just three flip-flops, B, C and D, are used. An output pulse at a frequency of 1 Hz is obtained from the output (pin 11) of flip-flop D.

Digital clocks and timing devices derive their 1 second pulses by dividing the mains frequency by 50 in this way, though different circuits may be used from the one described above. Accurate timing is possible since the mains frequency deviates very little from 50 Hz, though the mains voltage may vary considerably.

A further division by 6 and then by 10 gives one pulse per minute; division by 6 and 10 again, one pulse per hour; a further division by 6 and 4, one pulse per day. Of course, modern digital mains-operated clocks do not use individual BCD counters, and frequency division takes place within a purpose-designed *clock chip.*

6.2 Integrated circuit frequency dividers

The TTL 7493 and the CMOS 4516 counters each contain four cascaded flip-flops. Flip-flop A divides an input frequency by 2^1, flip-flop B by 2^2, flip-flop C by 2^3 and flip-flop D by 2^4. So the nth flip-flop of n cascaded flip-flops would divide an input frequency by 2^n. The more flip-flops there are, the lower the output frequency. Each output represents a bit, 0 or 1, of data which is binary weighted by powers of 2 — see Chapter 11. Thus the 7493 and 4516 are 4-bit counters, and the binary weights are 1, 2, 4 and 8.

Figure 6.4 shows one 12-bit and two 14-bit counters in 16-pin CMOS packages. All the bits are available from the 4040 so frequency division up to $2^{12} = 4096$ is possible. Obviously the 4020 and 4060 packages cannot provide all flip-flop outputs since there aren't enough pins on the package! The 4020 misses out division by 2^2, and 2^8, and the 4060 misses out division by 2^1, 2^2, 2^3 and 2^{11}. The 4060 provides maximum frequency division by a factor of $2^{14} = 2 \times 2 \times 4096 = 16\ 384$.

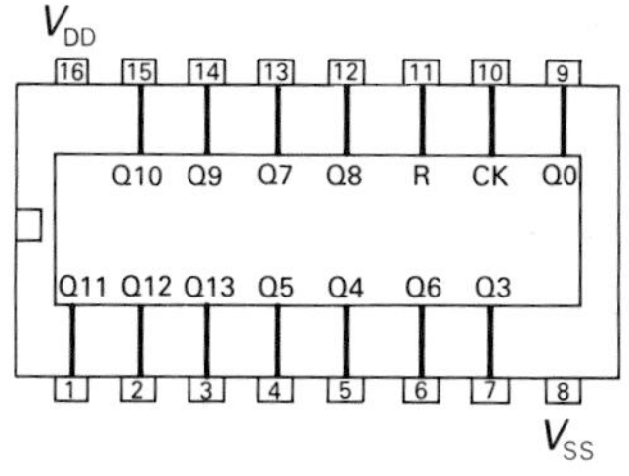

4020 14-bit counter

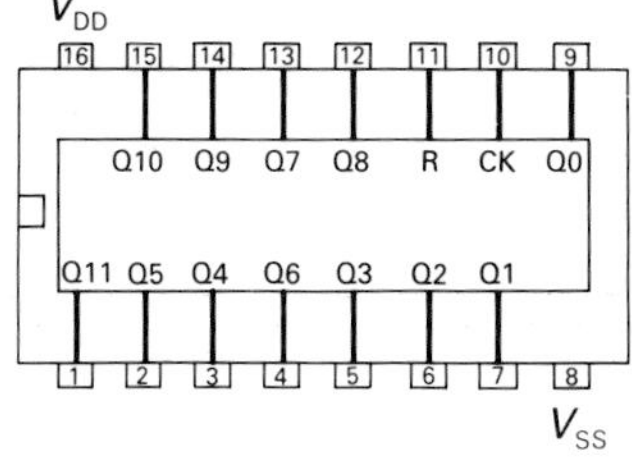

4040 12-bit counter

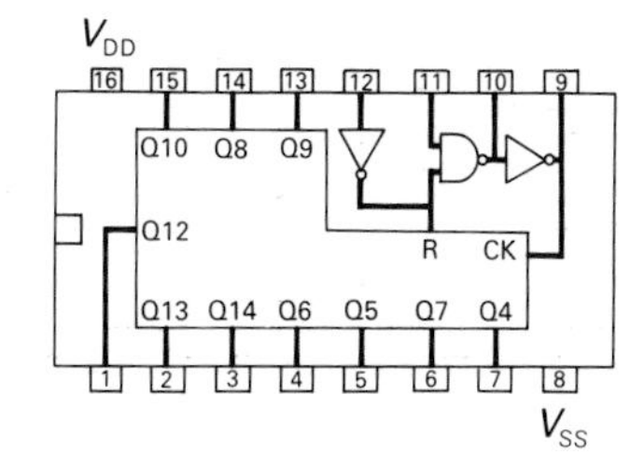

4060 14-bit counter

Figure 6.4 CMOS counter/dividers

These binary counter, or *frequency dividers*, are easy to use. The input signal, e.g. the pulses from a clock, must have rise and fall times faster than 5 microseconds (μs). This means that the signal must change from low to high and vice versa very quickly if the flip-flops are to be operated reliably. The output from a Schmitt provides this fast change, and the counter advances on the high-to-low transition (i.e. change) of the clock pulse. For counting to take place, the RESET input (e.g. pin 11 on the 4040) must be connected low, i.e. to 0 V. If the RESET is taken high, the counters reset to zero, i.e. all outputs at binary 0.

The 4060 14-bit counter is the basis of Project Module D5 which has an option for inputting the signal from a stable crystal oscillator — see Chapter 9. The frequency of the crystal oscillator is 32.768 kHz. Hence the lowest frequency obtained from pin 3 of the divider is

$$\frac{32\ 768}{2^{14}} = 2\ \text{Hz}$$

One extra flip-flop would divide this frequency by 2 and give a 1 Hz pulse for accurate timing purposes. The other outputs give higher frequencies, e.g. from pin 4, $32\ 768 \div 2^7 = 256$ Hz.

The 4020, 4040 and 4060 frequency dividers can divide an input frequency only by powers of 2. But the 4018 shown in figure 6.5 is a *presettable divide-by n* counter designed to divide by any whole number between 2 and 10. It contains five flip-flops and some complex logic circuitry.

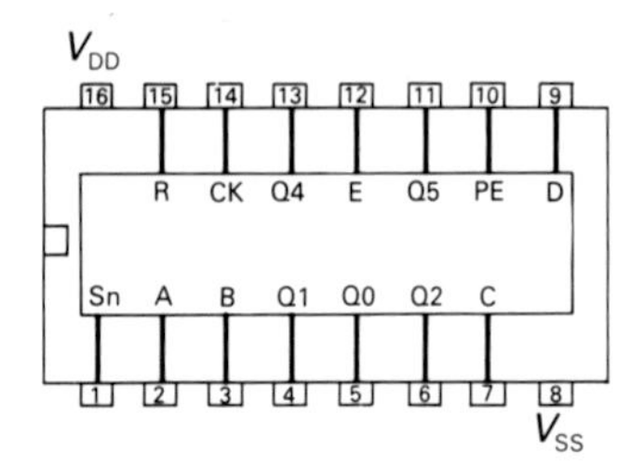

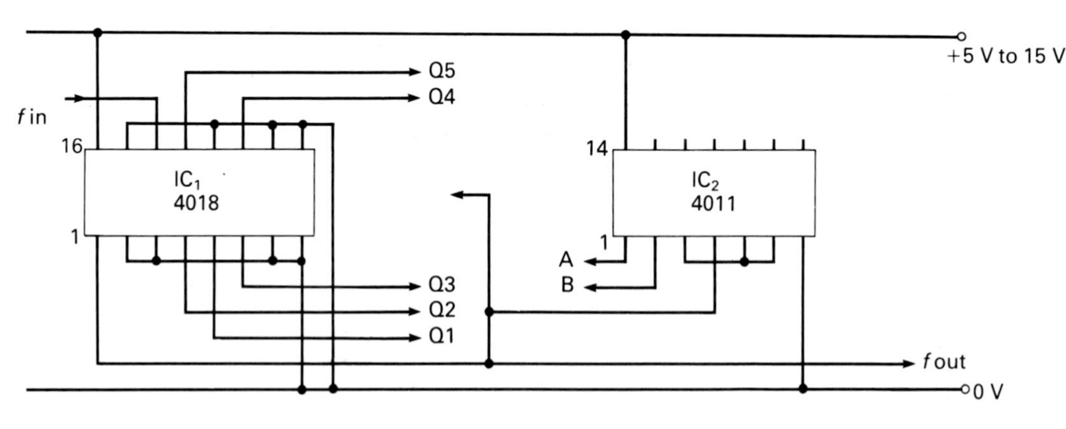

divide by	connect L to
2	Q1
4	Q2
6	Q3
8	Q4
10	Q5

divide by	connect A to	connect B to
3	Q1	Q2
5	Q2	Q3
7	Q3	Q4
9	Q4	Q5

Figure 6.5 Using the 4018 divide by *n* counter

If the data terminal (pin 1) is connected to pins 3, 5, 7, and 9, a frequency input to pin 14 is divided by 2, 4, 6, 8 and 10, respectively. Division by 3, 5, 7, and 9 requires the use of a 4011 quad 2-input NAND gate.

Any counter that divides an input frequency by ten is called a *decade counter*. Thus the 7490 and 4516 can be used as decade counters since they each have a divide-by-10 output. The 4017 device shown in figure 6.6 is also a decade counter but it doesn't have a binary output. Instead it has ten decimal outputs plus an overflow or CARRY OUT (CY) output. As the waveforms show, when its RESET (R, pin 15) and CLOCK ENABLE (Y, pin 13) terminals are at 0 V, these ten outputs go high in turn for successive low to high transitions of the clock pulse.

The circuit shows how to wire up a simple decimal 99 counter by cascading two 4017s. The CARRY OUT pulse from IC_1 (which measures the 'units' count) completes one cycle for every ten input pulses. This divide-by-ten pulse is used to clock IC_2 which measures the 'tens' count. By momentarily bringing the RESET high by means of SW_1, all outputs are set low except output '0' (pin 3) which remains high. Twenty LEDs with 1 kΩ series resistors show the count at the outputs. A third 4017 could be used to count up to 999.

The 4017 makes a cheaper decimal counter than one using seven-segment displays — see Chapters 2 and 7. The 4017 forms the basis of the Decade Counter (Project Module C3) where its use for control and counting is explained, as well as how to operate the counter to divide an input frequency by any number from 2 to 9. The 4017 is also useful as a multiplexer (Chapter 8) and is used in the Data Recorder (Section 17.8).

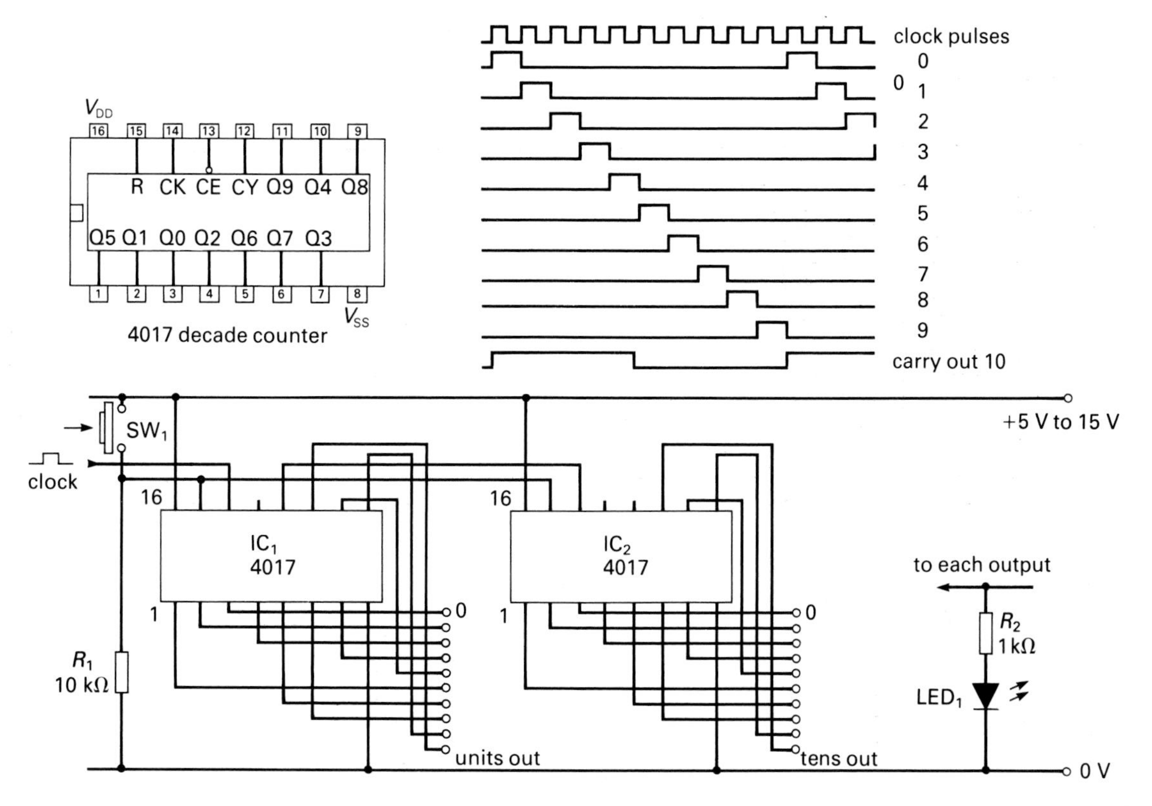

Figure 6.6 Using the 4017 decade counter

7 Encoders and Decoders

7.1 Introduction

Encoders and decoders are generally known as code converters for they have the useful function of changing information from one form to another. As shown in figure 7.1, an encoder is an input device and a decoder is an output device:
An *encoder* changes the format of information entered into a system.
A *decoder* changes the format of information taken from a system.

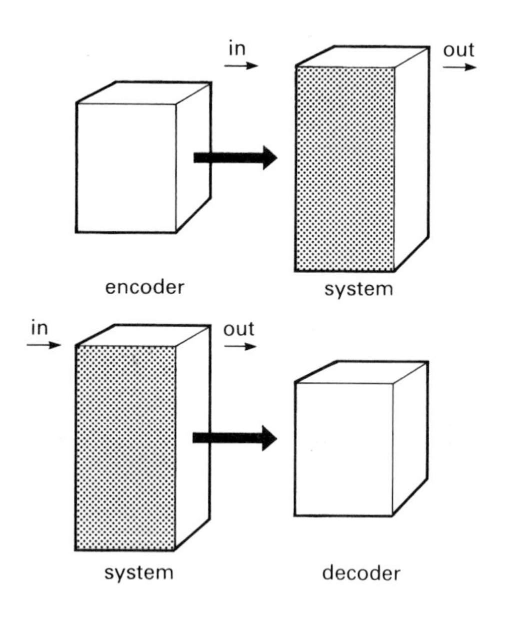

Figure 7.1 Encoders are input devices; decoders are output devices

It is useful to think of an encoder as a device which enables a digital system to recognise readily understood decimal data. For example, a keypad on a security system is a type of encoder. It converts a decimal number input to the security system, into a binary code. This code is then processed by the circuits of the security system which decide whether the correct sequence of decimal numbers has been entered.

It is useful to think of a decoder as a device which extracts useful information from a digital system. For example, BCD-to-seven-segment decoder drivers are used in digital watches. They convert digital information output from the watch circuits into a pattern of signals for displaying readily understood decimal numbers on a seven-segment display.

Encoders and decoders make use of combinational logic devices, and it is quite possible to construct them from individual logic gates. However, it is easier to use purpose designed integrated circuit packages, some examples of which are given in the following sections.

7.2 A keyboard encoder

Figure 7.2 shows how simple it is to design a keyboard encoder using the 74C922 IC which is housed in an 18-pin DIL package. This CMOS device operates from a supply voltage in the range 3–15 V. The circuit fully encodes an array of 16 normally-open single-pole single-throw push switches, SW_1 to SW_{16}. Thus when SW_4 is pressed and released, the encoder produces a 4-bit binary output of (0111). This output remains until the next key is pressed. Just like the much more complex keyboard of a microcomputer, the 16 switches are scanned one after the other by an oscillator whose frequency is determined by capacitor C_1. The light-emitting diodes, LED_1 to LED_4, indicate the value of the binary output converted by the keyboard encoder. This keyboard is the Project Module E3 described in this book.

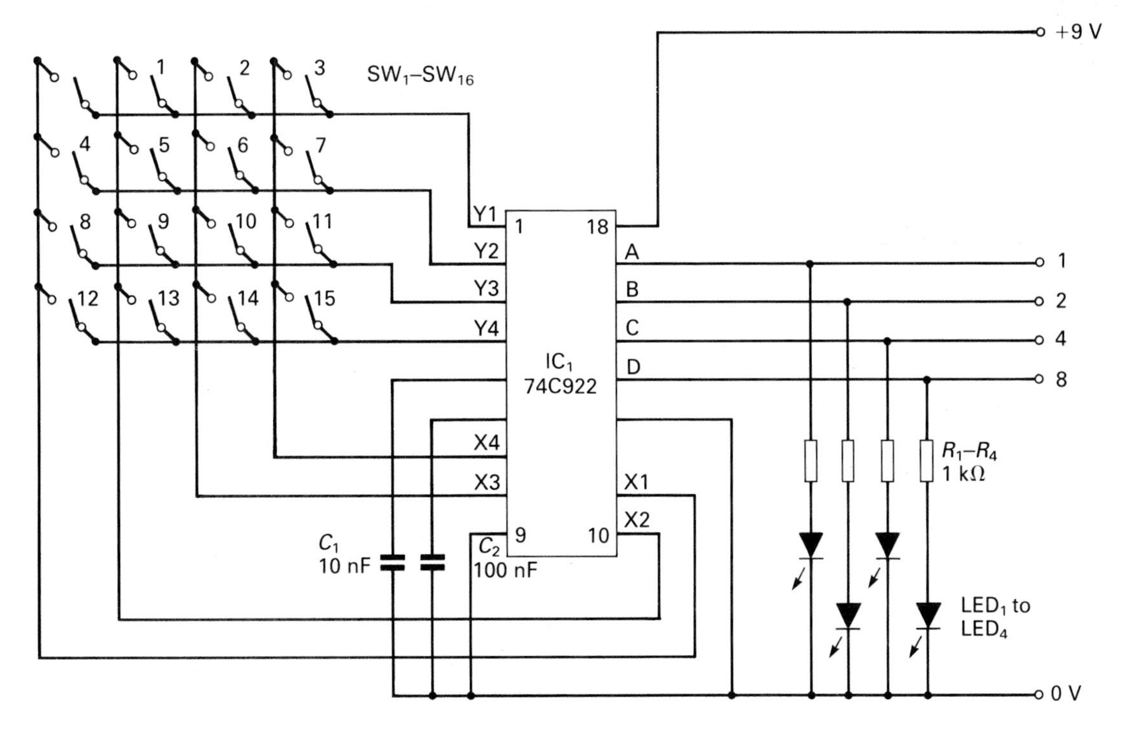

Figure 7.2 Circuit diagram of the keyboard encoder

7.3 BCD-to-decimal decoders

The TTL and CMOS IC packages shown in figure 7.3 accept a 4-bit BCD code and produce a decimal output. The 4028 CMOS device is the basis of Project Module E5. If this module were used with Project Module E6 (Infrared Remote Control), the first ten states of the 4-bit binary output of the infrared receiver could be decoded into ten discrete channels for control purposes.

Most decoders contain from 12 to 50 gates and are therefore examples of *medium scale integration (MSI)* circuits. Figure 7.4 shows how ten 4-input NAND gates and four inverters are used in the 7442 BCD-to-decimal package.

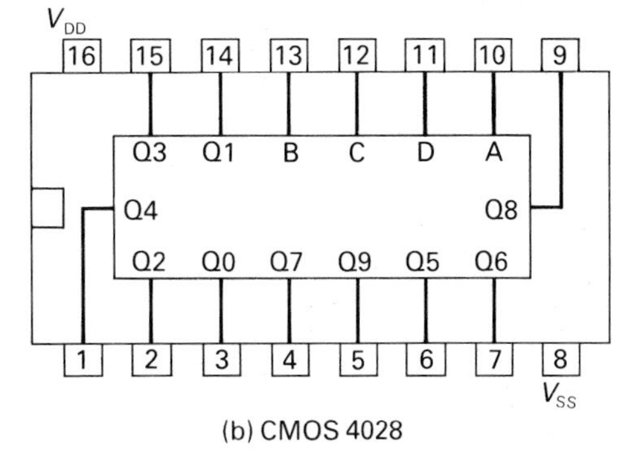

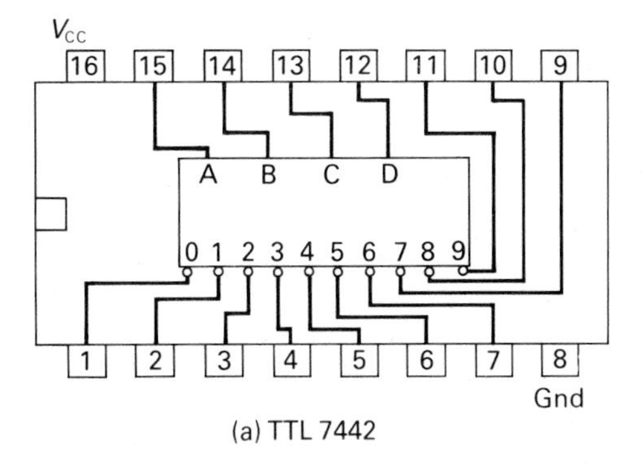

Figure 7.3 Two types of BCD-to-decimal decoders

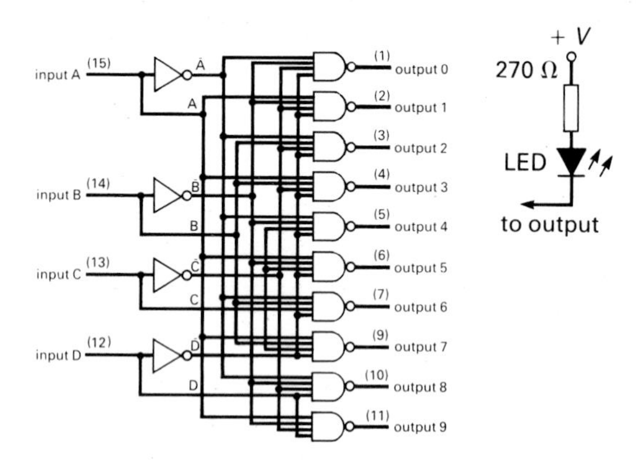

Figure 7.4 Logic diagram of a BCD-to-decimal decoder

The truth table in figure 7.5 lists the input and output signals for the 7442.

Figure 7.5

BCD inputs				decimal outputs								
D	**C**	**B**	**A**	**9**	**8**	**7**	**6**	**5**	**4**	**3**	**2**	**1**
0	0	0	0	1	1	1	1	1	1	1	1	1
0	0	0	1	1	1	1	1	1	1	1	1	0
0	0	1	0	1	1	1	1	1	1	1	0	1
0	0	1	1	1	1	1	1	1	1	0	1	1
0	1	0	0	1	1	1	1	1	0	1	1	1
0	1	0	1	1	1	1	1	0	1	1	1	1
0	1	1	0	1	1	1	0	1	1	1	1	1
0	1	1	1	1	1	0	1	1	1	1	1	1
1	0	0	0	1	0	1	1	1	1	1	1	1
1	0	0	1	0	1	1	1	1	1	1	1	1

Note that the outputs are 'active low'. This means that output voltages are normally high but go low when a BCD input has been decoded. The 7442 thus *sinks* current at its output and LEDs must be connected between the output and the + 5 V supply as shown in figure 7.4 if they are to light. The outputs from the 4028 are 'active high' so that it *sources* current at its output. Thus LEDs must be connected between the output and 0 V if they are to be lit by the high signals at the output. For the 4028, the 1s and 0s are interchanged for the decimal outputs in the truth table shown above.

7.4 BCD-to-seven-segment decoders

Each of the IC packages shown in figure 7.6 accepts a 4-bit code from a BCD counter and delivers decoded digital signals, i.e. 1s and 0s, for operating the LEDs in seven-segment LED displays. These are known as *seven-segment decoder/drivers* since they provide sufficient current to light the LEDs in the display.

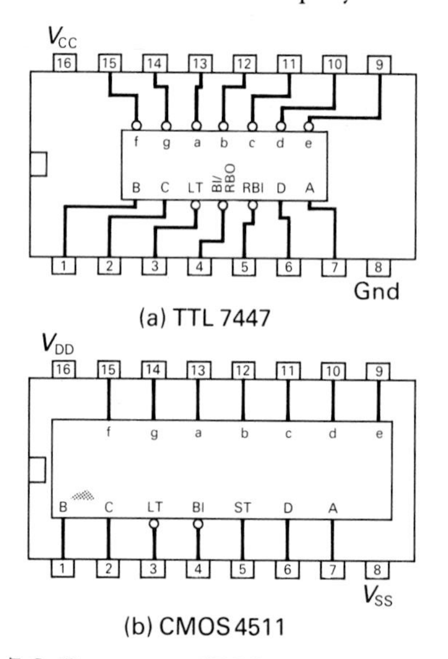

Figure 7.6 Two types of BCD-to-seven-segment decoders

As shown in Section 2.2, the seven segments are labelled *a* to *g*. Figure 7.7 shows the connections required for each type of decoder to light these segments. Note that outputs from the 7447 are 'active low'. For example, outputs *a*, *b*, *d*, *e* and *g* have to go low to light the corresponding segments and produce the number 2. This means that the seven-segment display must be a common-anode type so that the cathode of a light emitting diode is grounded to light that LED. Thus the 7447 sinks current (up to 40 mA per segment) into its output terminal when it is low.

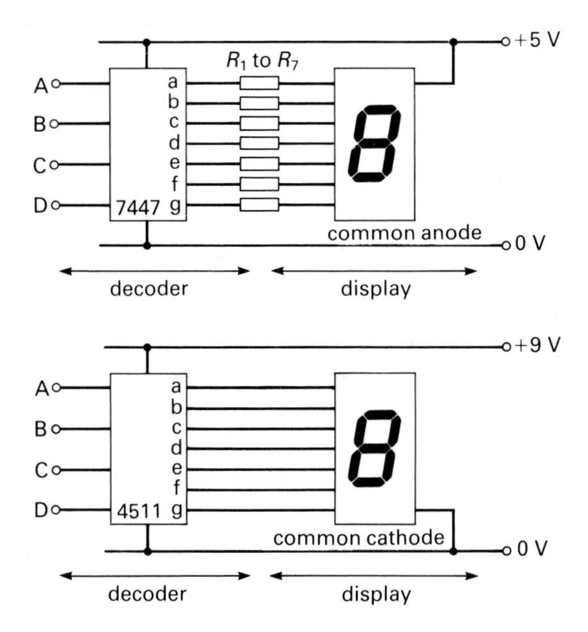

Figure 7.7 Principal connections for the 7447 and 4511 decoders

The CMOS 4511 decoder is designed to operate a common-anode display since its outputs are 'active high', i.e. it sources current (up to 5 mA per segment at 10 V supply). Note that series resistors must be provided between the 7447 and its seven-segment display, but not between the ~~4026~~ 4511 and its display provided a supply voltage of less than 10 V is used.

Two BCD counters and two decoders are required for a two-digit display, the BCD counters being cascaded as explained in Section 5.7. The CMOS 4026 counter/decoder/driver contains a BCD counter and a seven-segment decoder in a single 16-pin package, and that makes it easy to assemble multi-digit displays. Also, using a 9 V supply the series resistors are not needed for the segments of the common cathode display. Project Module D7 shows how to make a two-digit decimal counter using the 4026 device.

The 7447 produces the pattern of 1s and 0s to display the decimal numbers 0 to 9, shown in figure 7.8. Note that this truth table is valid if the LT, BI/RBO and RBI pins are held high as explained in Section 7.5. Note too that the TTL 7448 device is designed to operate a common-cathode display, i.e. its outputs are active high like the 4511.

7.5 Experiment E7

Using the 7447 decoder/driver

Figure 7.9 shows how you might look into the operation of the 7447 decoder/driver by assembling a circuit on breadboard. The 7490 BCD counter is clocked by input pulses from a Schmitt trigger oscillator as

Figure 7.8

BCD input D	C	B	A	segment outputs a	b	c	d	e	f	g	decimal number
0	0	0	0	0	0	0	0	0	0	1	0
0	0	0	1	1	0	0	1	1	1	1	1
0	0	1	0	0	0	1	0	0	1	0	2
0	0	1	1	0	0	0	0	1	1	0	3
0	1	0	0	1	0	0	1	1	0	0	4
0	1	0	1	0	1	0	0	1	0	0	5
0	1	1	0	1	1	0	0	0	0	0	6
0	1	1	1	0	0	0	1	1	1	1	7
1	0	0	0	0	0	0	0	0	0	0	8
1	0	0	1	0	0	0	1	1	0	0	9

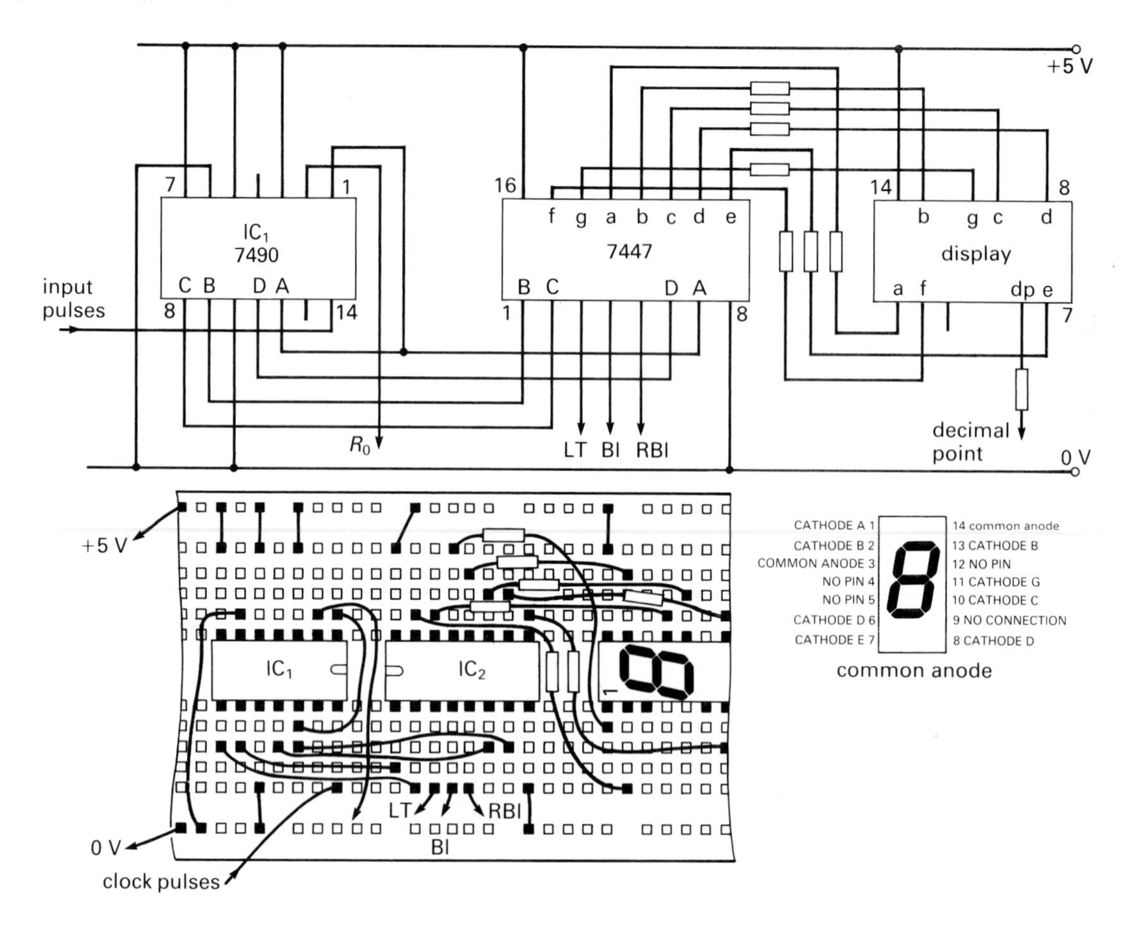

Figure 7.9 Experiment E7: Using the 7447 decoder/driver

in figure 5.3. The 'reset to zero' pin, R0, is connected to 0 V for counting to begin and removed to reset the DCBA outputs to $0000_{(2)}$.

The DCBA outputs from the 7490 are connected to the DCBA inputs of the 7447. The *a* to *f* outputs of the 7447 are connected to the appropriate segments on the common-anode seven-segment display via 270 Ω series resistors.

The display will count 0 to 9 and then back to 0 if the following connections are made:

On the 7490, connect R0 to 0 V.
On the 7447, allow LT (lamp test), RBI (ripple-blanking input), and BI/RBO (blanking input/ripple-blanking output) to *float*, i.e. take a logic high.

Note the following effects of connecting the LT, and RBI pins to 0 V:

Connect LT (pin 3) to 0 V and all segments will light; this is a test facility to check that all segments are working.
Allow LT to float high and connect RBI to 0 V; watch the display and you will see that decimal 0 is not displayed though it is still counted in the sequence.

▽ 7.6 Leading zero blanking

The 7447's facility for blanking 0s on the display is useful in multi-digit displays, and can be explained by means of the two-digit counter in figure 7.10 which counts from 0 to 99. Pulses arriving at the input of

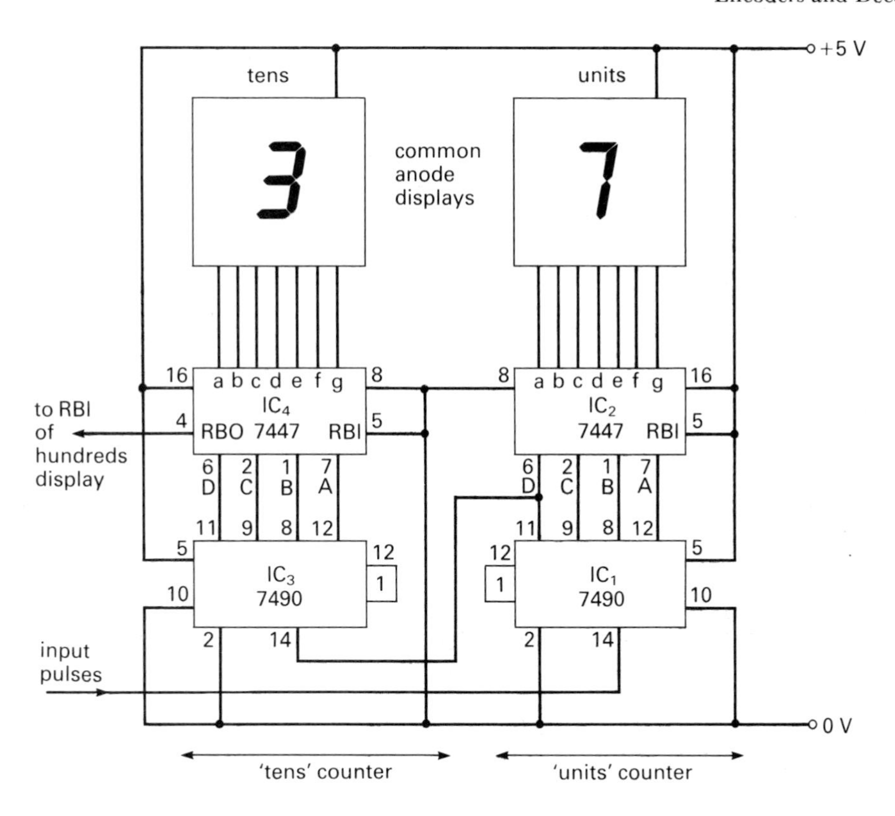

Figure 7.10 Two-digit counter using TTL devices and featuring leading zero blanking

the 'units' counter, IC_1, are counted to produce a BCD output which is decoded by IC_2 to produce the decimal 'units' count. The Q_D output from IC_2 is fed to the Q_A input of IC_3. When Q_D goes low as the input from IC_2 changes from binary 1001 to binary 0000, IC_3 registers one pulse. Thus IC_3 counts the 'tens' pulses.

The ripple-blanking input (RBI) of IC_2 is connected to + V, i.e. it is high so that the units digit indicates a zero as required, e.g. for 20, 40, etc. But the RBI pin of IC_4 is connected to 0 V, i.e. it is low, so that whenever this decoder receives a 0000 signal its outputs all go high and the display is blanked. Thus the 'tens' digit is blank for counts from 0 to 9, and shows 1 for counts 10 to 19. The blanking of the 'tens' zero display in this way is known as *leading zero blanking* (l.z.b.).

The ripple-blanking output (RBO) of the 7447 is used if the display system has more than two digits. Its operation depends on the fact that the conditions which produce a blank display, i.e. 0000 input and RBI low, will cause RBO to go low. Thus by wiring the RBO output of the 'tens' decoder to the RBI input of the 'hundreds' decoder, the 'hundreds' digit will show a zero when the 'thousands' digit shows 1 to 9; e.g. the display will show △ 6024 rather than 6(blank)24.

▽ 7·7 Driving liquid crystal displays

The general structure of a liquid crystal display (LCD) is described in Section 2.3. In an LCD, each display segment relay allows light to pass through it, or blocks the light, so the LCD is not self-illuminating like the seven-segment LED display.

To maintain contrast, and to give the LCD a good life expectancy, the LCD has to have special drive circuitry to make sure that there is no d.c. signal across the display segments. In the example shown in figure 7.11, the binary number 1001 is being received by the BCD decoder, e.g. the CMOS 4543 device. The decoder therefore activates the *a*, *b*, *c*, *f* and *g* outputs (the decoder outputs are active high). The *d* and *e* outputs are low. The backplane of the display receives a 30 Hz square wave signal which is also applied to one input of each of the CMOS exclusive-OR gates used to drive the LCD segments.

These XOR gates ensure that when the segment inputs are high, the segment drive voltage and the backplane voltage are exactly 180° out of phase. This is shown in figure 7.11 by the waveforms being the inverse of each other. Thus there is an overall a.c. voltage across *a*, *b*, *c*, *d* and *g* segments which results in these segments being dark. Where the segment input is low, the segment input and the backplane are exactly in phase: there is no longer a voltage across the liquid crystal and the △ segment remains transparent.

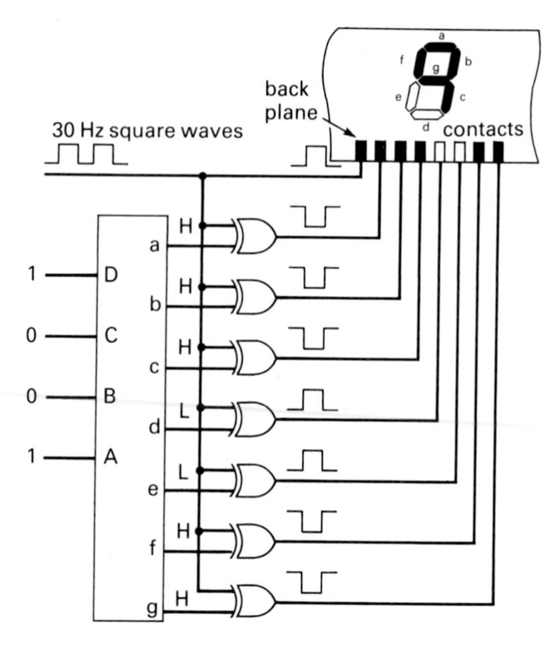

Figure 7.11 Basic drive circuit for LCD displays

8 Multiplexers and Demultiplexers

8.1 Introduction

A *multiplexer* is a device for selecting a single piece of data from several pieces of data. For example, the channel selector switch on a television is a multiplexer, or a *data selector*. The reason for using multiplexers in electronic systems is to cut down on the amount of circuitry and interconnections within the system.

A multiplexer is much like a rotary single-pole multiway switch. Figure 8.1 shows how a 4-input multiplexer is made from a single-pole 4-way switch. The positions of the switch determine which piece of data, D_0 to D_3, is passed to the output; the single output line carries any one of the four pieces of data one piece at a time, i.e. serially. The rotary switch acts as a data selector or multiplexer.

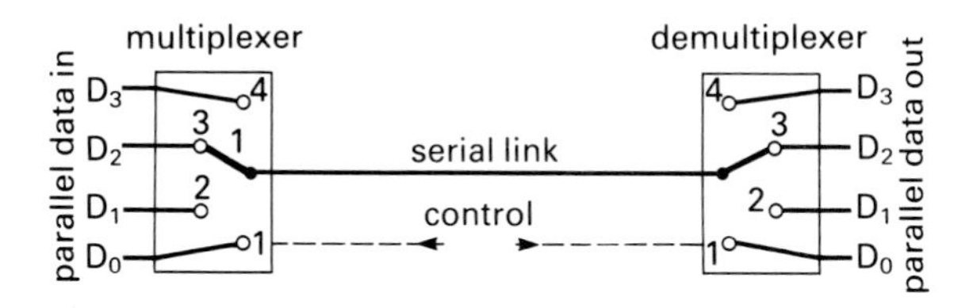

Figure 8.1 Using rotary switches as multiplexers and demultiplexers

The transmitted data can be demultiplexed at the other end of the transmission line by using a second single-pole 4-way switch. Thus if the switches are in position 3, the data, D_2, on input 3 is routed to output 3. If the switches are synchronised and rotated rapidly by the control circuit, a single line connecting multiplexer and demultiplexer carries all four pieces of data which share the common connection.

▽ 8.2 Data transmission

Figure 8.2 shows the principle of using a multiplexer to convert an 8-bit data word from parallel to serial form for transmission along a single wire.

The binary data word is latched on the inputs, D_0 to D_7, of the multiplexer, MUX_1. A clock operates a 3-bit binary counter which provides a select code, CBA, which decides the bit of data to be sent to the output line. As the clock advances by one pulse, the bit of data on the next input line is selected by the counter and sent to the output. After eight clock pulses, the whole 8-bit data word has been sent, one bit at a time to the output and now exists in serial form. This 1-of-8 multiplexer thus behaves like a single-pole 8-position rotary switch with the switch position controlled by the state of the three select inputs, A, B and C.

In order to recover the 8-bit word at the end of the transmission line, a demultiplexer, $DEMUX_1$, like that shown in figure 8.3 is required. Serial data input to $DEMUX_1$ is synchronised to the input clock and to the code on the select lines. As the binary count on the select lines changes, the input data stream is transferred to the output of $DEMUX_1$, one bit at a time. After eight clock pulses, the whole data word is *recaptured* and appears at the output of the demultiplexer.

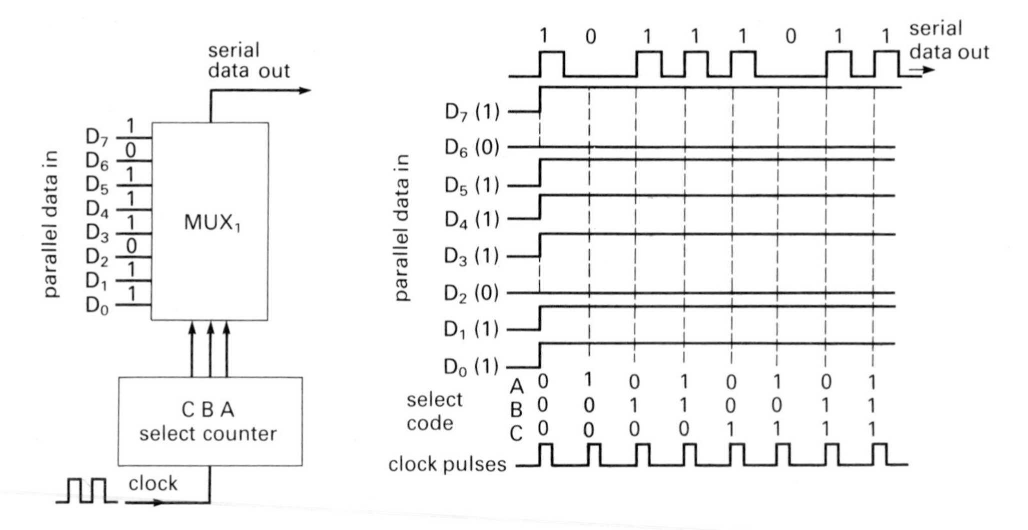

Figure 8.2 8-input multiplexer action

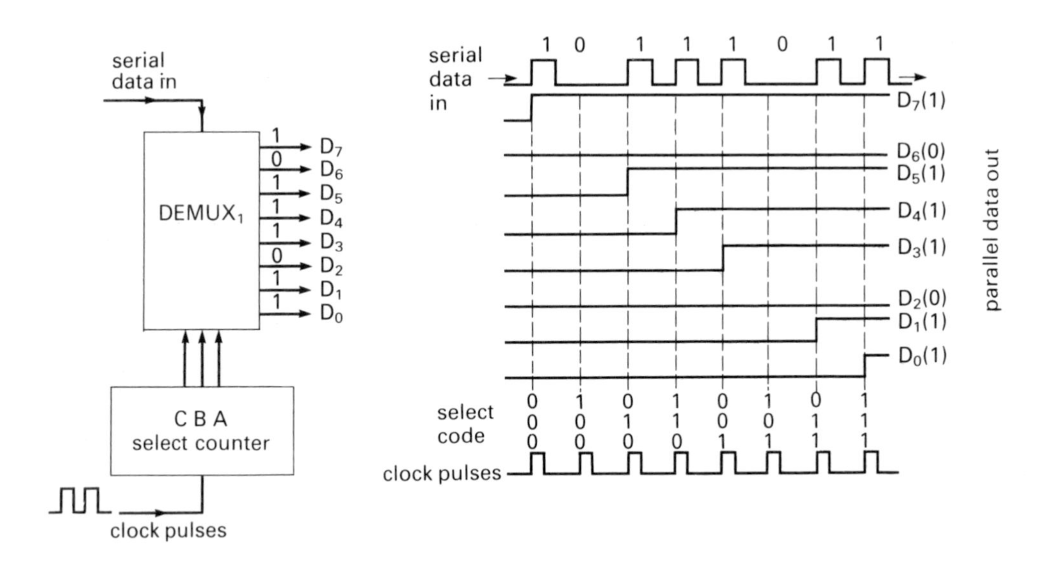

Figure 8.3 8-output demultiplexer action

Many data processing systems, including computers and optical fibre communications systems, make use of multiplexers and demultiplexers to transmit data as a stream of 0s and 1s over communications links. A good example of how a binary signal can be transmitted as a series of on/off pulses of infrared light, and then be recaptured as a complete data word at the receiver, is demonstrated by Project Module E6 (Infrared Remote △ Control) described in Chapter 18.

▽ 8.3 A practical data transmission system

Figure 8.4 shows a practical data transmission system which makes use of the principles described above in Section 8.2.

The data to be transmitted is placed on the 16 inputs of the 74150 multiplexer, IC_1. This multiplexer has an enable input (pin 9) which is connected to 0 V to enable it to transfer data to its output (pin 10). This data travels along the serial data link to the input (pin 19) of the 74154 demultiplexer, IC_2, where it is routed to a particular output. Normally the output of IC_2 is inverted as shown by the small circles. Thus 7404 hex inverters are used to *complement* the output to the original data. The transfer of data from the multiplexer to the demultiplexer one bit at a time is controlled by the 7493 4-bit counter, IC_6.

When the 4-bit control signal, Q_D to Q_A, of the 7493 counter is at 0000, the data on the D_0 input of IC_1 is transferred, via the serial link, to the D_0 output of IC_2. On the next clock pulse, when the 7493 output is at 0001, the data on the D_1 input is transferred to the D_1 output of IC_2. The counter continues to scan the inputs of the multiplexer and to transfer this data, simultaneously, to the output of the △ demultiplexer.

8.4 Multiplexing seven-segment displays

If we want to extend the two-digit display shown in figure 7.7 to three-four or more digits, the circuit becomes complicated; each digit requires its own decoder/driver and all the digits are on at the same time. However, if the display is multiplexed, all the digits share one decoder/driver which

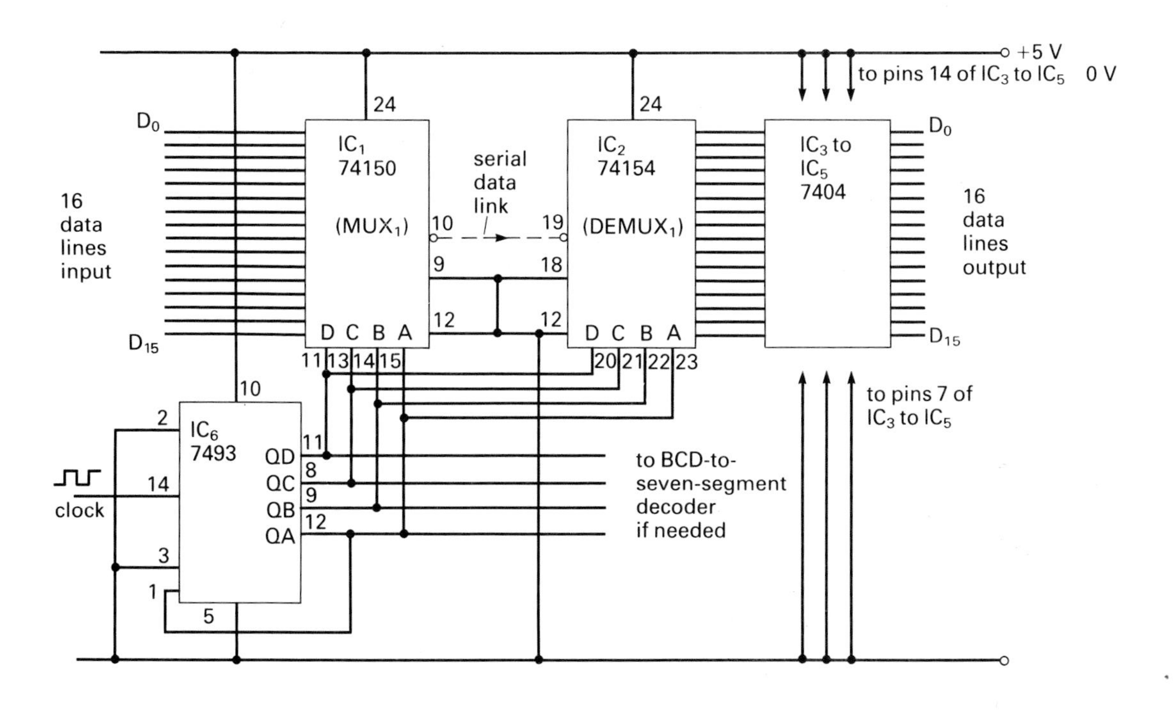

Figure 8.4 A practical 16-bit data transmission system

reduces circuit connections and power consumption.

Multiplexing scans the digits rapidly one after the other so that they all appear to be on due to the sensitivity of the eye; i.e. flashes of light which occur at a frequency above 20 to 30 Hz 'run together' to give the effect of a continuous glow. A multiplexed display illuminates all the appropriate segments of each digit for a fraction of second, and immediately after does the same for next one. Figure 8.5 shows how this is done for a multiplexed 4-digit seven-segment display.

The scanning oscillator synchronises the two multiplexers, MUX_1 and MUX_2, which act like single-pole 4-way rotary switches. Under the control of the clock, MUX_1 takes the BCD data from each BCD counter in turn and routes it to the decoder/driver. At the instant that the BCD data, e.g. 0011, from the units counter is sent by MUX_1 to the decoder/driver, the least significant digit in the display is selected by MUX_2 which energises that digit by connecting the anodes of the LEDs of that display to + V (assuming they are common-anode devices). Thus the number 3 is displayed. On the next count of the clock, the BCD data, e.g. 0111, from the 'tens' counter is routed to all the digits by MUX_1. At the same time, MUX_2 selects the 'tens' digit to display the number 7 as required. On the next clock pulse, the 'hundreds' digit is activated by MUX_2 as the BCD data 1001 is routed to MUX_1. The thousands digit is activated in the same way to display the number 1. Thus the number '1973' appears on the display. A clock frequency of about 100 Hz produces a flicker-free display.

Note that only one decoder/driver is used regardless of the number of digits in the display. Also the power consumption is reduced since the digits are on intermittently, and wiring is simpler.

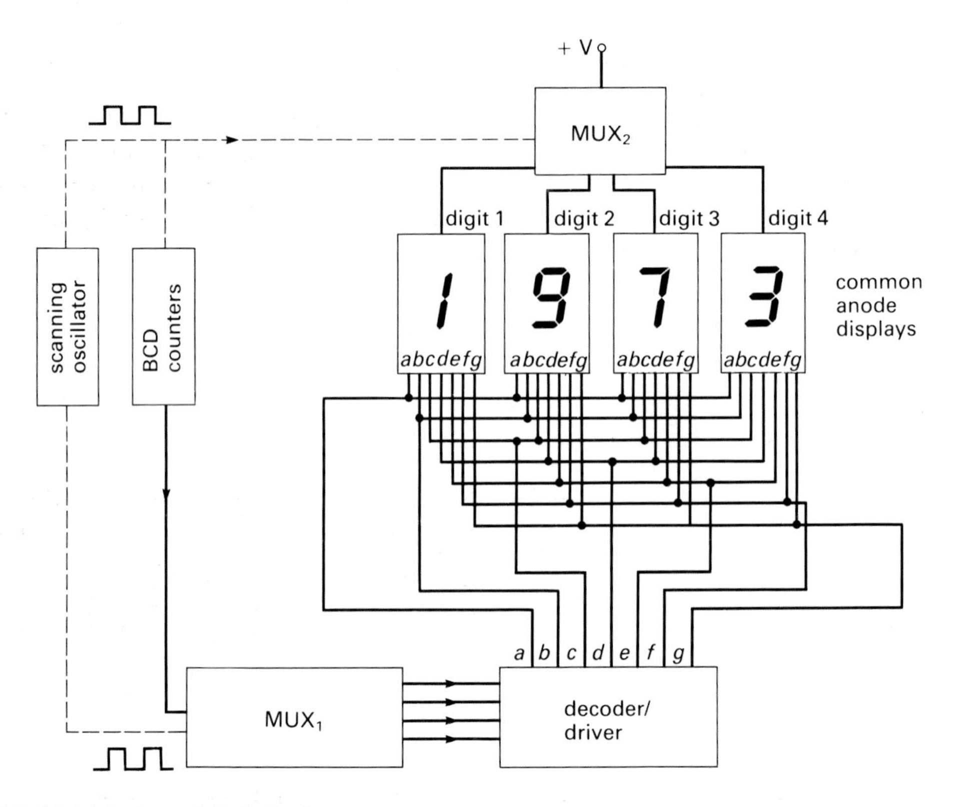

Figure 8.5 Multiplexing a 4-digit display

The brightness of the display can be varied by varying the period during which each display is off. This is usually achieved by varying the mark-to-space ratio of the clock which operates the multiplexers. Since each segment LED is only illuminated for a short time, fairly high maximum currents can be passed through each segment resulting in a bright display, yet one which requires, on average, less current than a non-multiplexed one. All the circuitry like the counters, decoders, and multiplexers, which are required to drive four or more digits, are readily available in integrated circuit packages for clocks, frequency counters, etc.

Demultiplexers are sometimes called decoders. Thus the CMOS 4017 decade counter (Section 6.2) and the TTL 7447 seven-segment decoder/driver (Section 7.4) are types of demultiplexers.

8.5 Analogue switches

There is one thing CMOS devices can do that TTL cannot: they can switch analogue signals on and off using digital signals.

Figure 8.6(a) shows that an *analogue switch* behaves like a push-to-make release-to-break switch, SW_1, which is controlled by a signal on the control input, C. A logic 1 on the control input closes the switch and V_{in} appears at output A. A logic 0 keeps the switch open. Analogue switches are also known as *transmission gates* for they allow the transmission of signals between input and output when the switch closes. Note also that analogue switches are bi-directional, i.e. signals can pass in either direction through the device.

The 4066 CMOS device shown in figure 8.6(b) contains four analogue switches, or transmission gates. Each gate has its control input. Thus as control input, Ca, is taken to logic 1, switch SW_1 is closed and a low resistance path of about 100 Ω is established between the input A and output A terminals of the gate. Before the logic 1 signal is applied, the resistance between the input and output is extremely high being about 10 TΩ (10^{13}) ohms.

Figure 8.7 shows how to make a *programmable-gain amplifier* using the CMOS 4066 device which contains four analogue switches. The gain is selected by taking the required gain-select pin high. This closes the switch and connects the corresponding feedback resistor, R_1 to R_4 between the inverting and output pins of the op amp. In this way, four values of voltage gain can be selected. This circuit is a non-inverting amplifier and is based on an op amp such as the 741 or the 3140. See Book D, Chapters 2 to 5, for information about how op amps are used in this way.

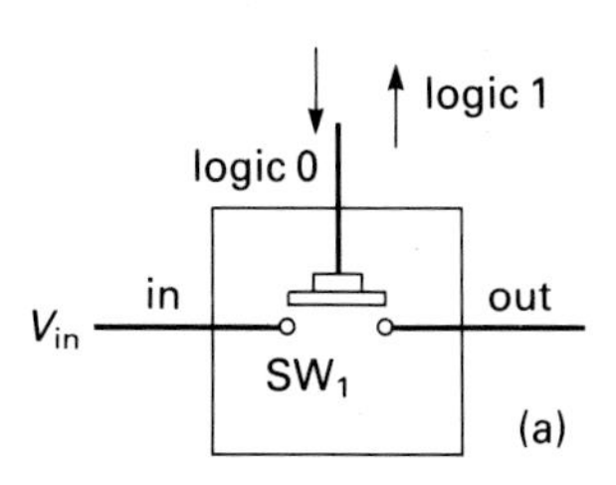

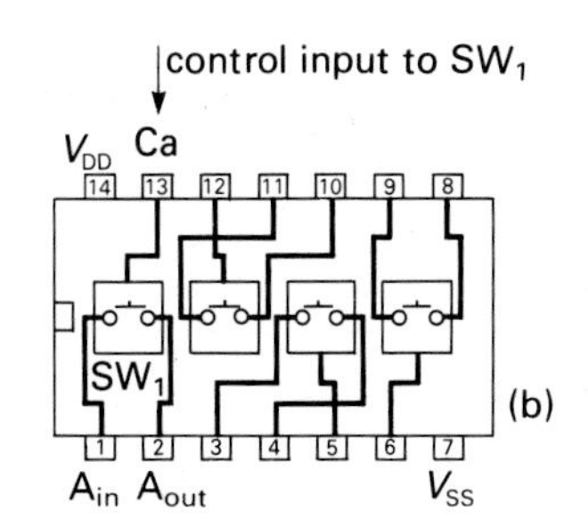

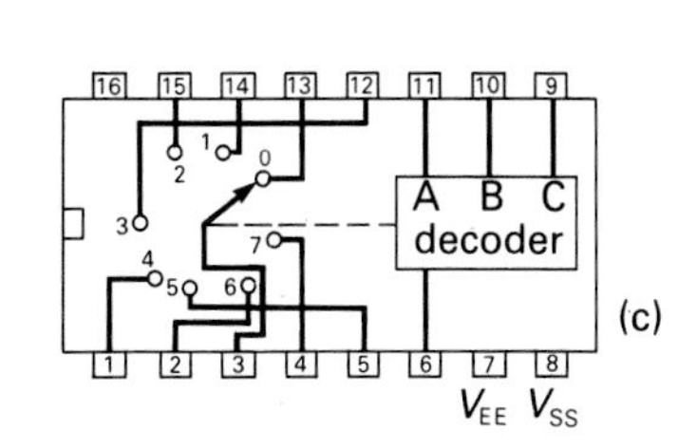

Figure 8.6 (a) The principle of an analogue switch (b) The CMOS 4066 quad analogue switch (c) The CMOS 4051 1-to-8 analogue switch

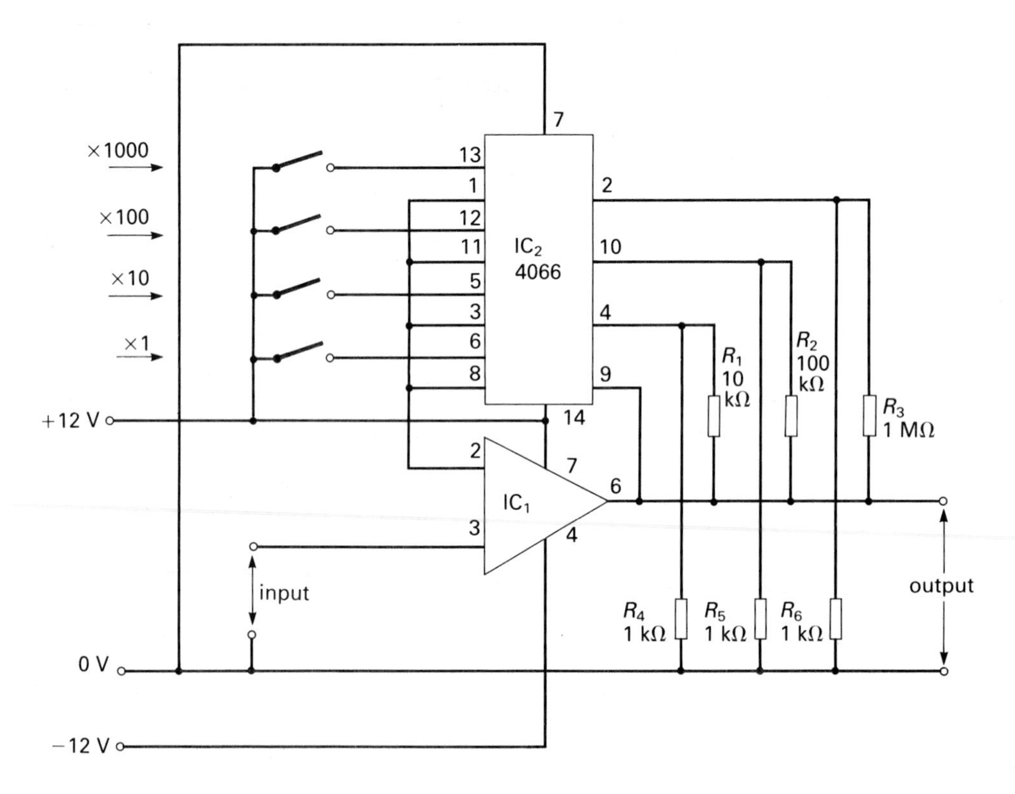

Figure 8.7 Programmable gain amplifier using the CMOS 4066 quad analogue switch

Figure 8.6(c) shows the CMOS 4051 1-to-8 analogue switch. This device has three select inputs, A, B and C. A 3-bit binary code on these inputs connects the input (pin 6) to one of the output switches (0 to 7) according to the truth table (figure 8.8).
Because it behaves in this way, the 4051 is of use as an analogue multiplexer/demultiplexer. Note that if pin 7 is taken to − 5 V, analogue signals between + 5 V and − 5 V can be controlled. There is an application of the 4051 in Section 17.9; it's a simple 'music box' in which the 4051 switches in different resistors in an astable circuit based on a 555 timer.

Figure 8.8

select code			**pin 6 (inhibit)**	**output enabled**
C	B	A	1	all channels off
0	0	0	0	0
0	0	1	0	1
0	1	0	0	2
0	1	1	0	3
1	0	0	0	4
1	0	1	0	5
1	1	0	0	6
1	1	1	0	7

9 Schmitt Triggers, Astables and Monostables

9.1 Introduction

The Schmitt trigger is a snap-action electronic switch which goes off and on at two specific input voltages, the upper and lower *threshold voltages*. As you will see the Schmitt trigger is widely used to 'sharpen up' slowly changing waveforms to provide inputs to digital systems, and for eliminating noise in circuits.

Figure 9.1 shows what a Schmitt trigger does to a smoothly changing sine wave at its input; it changes it into an abruptly changing square wave at its output. Notice that the Schmitt trigger acts as simple analogue-to-digital converter (ADC) — the ADC is discussed in Book D, Chapter 10.

In order to bring about the change to the input signal shown in figure 9.1, the Schmitt trigger has two threshold voltages, an upper threshold voltage, V_2, and a lower threshold voltage, V_1. Assume the input voltage is rising from 0 V. In this condition, the output voltage is high. When the input voltage passes the upper threshold voltage, V_2, there is a rapid fall in the output voltage to a low state. The output remains low while the input voltage continues to rise and fall again. However, the output voltage does not go high again until the input voltage has fallen to the lower threshold voltage, V_1. Note that the output voltage has only two values, high and low.

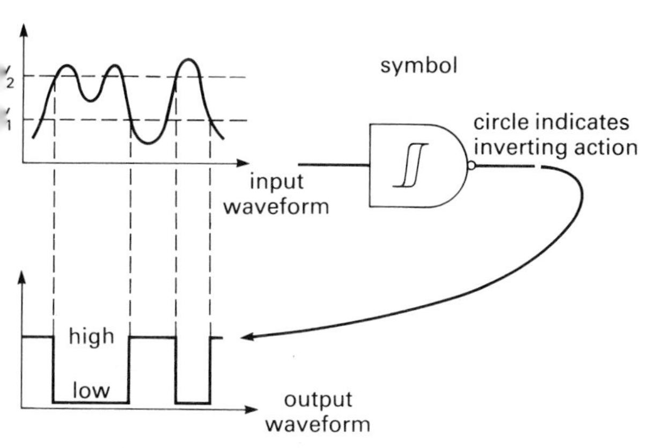

Figure 9.1 Schmitt trigger action

Thus the Schmitt trigger possesses *hysteresis* or *backlash*, i.e. it has a different switching threshold for increasing and decreasing voltages. The difference between the upper and lower threshold voltages is the hysteresis, V_h, of the circuit, i.e. $V_h = V_2 - V_1$. The fat S-shaped symbol inside the circuit symbol for a Schmitt trigger indicates the backlash effect of the circuit. The S-shape actually symbolises the shape of a hysteresis curve obtained when ferrous materials are magnetised and elastic materials are strained. Though in the case of the Schmitt trigger, a box-shape accurately describes the rapid rise and fall of the output voltage. The Schmitt trigger action is caused by positive feedback between the output and input, and this is best explained using an operational amplifier — see Book D, Section 4.6.

9.2 Schmitt trigger IC packages

Figure 9.2 shows Schmitt trigger packages in the TTL and CMOS range. These are inverting Schmitt triggers, i.e. the output

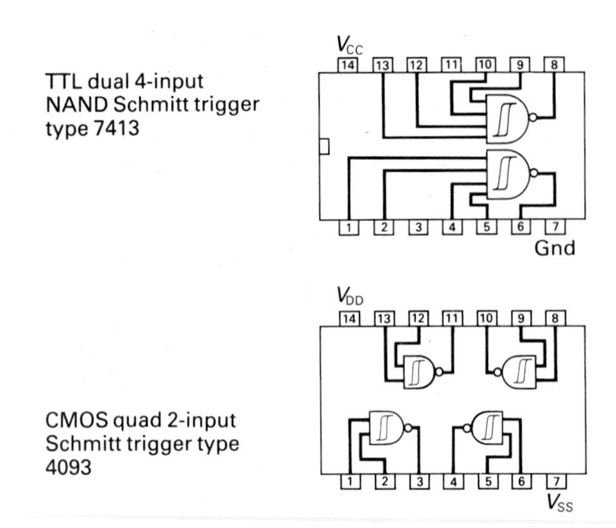

Figure 9.2 Two further Schmitt trigger packages (see also figure 5.2)

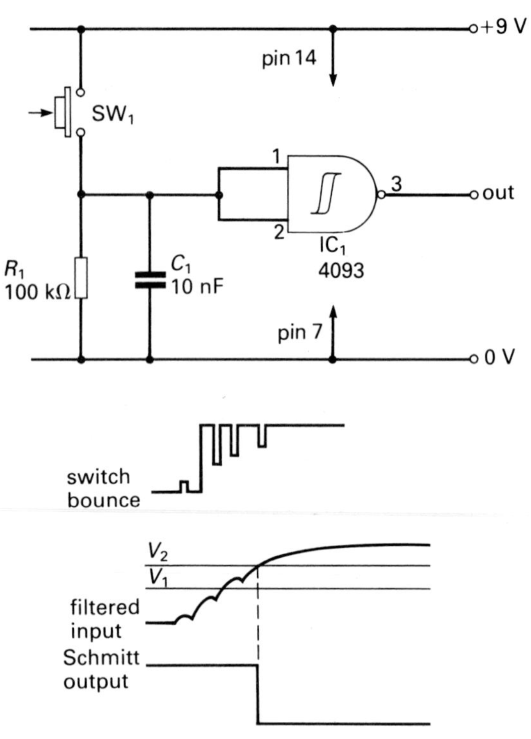

Figure 9.3 Switch debouncer

is high when the input is low, just like the TTL 7414 and the CMOS 40106 'hex' packages used to provide clock pulses to operate binary counter circuits in Chapter 5 — see figures 5.2 and 5.3.

9.3 Switch debouncer and clock

Figure 1.5 showed how to use a Schmitt trigger as a switch debouncer and as a clock, i.e. an oscillator or astable. You should connect all the inputs together if you use the TTL 7413 or the 4093.

Bounce of the mechanical contacts of a switch can generate multiple pulses, only one of which is intended. Figure 9.3 shows the input and output pulses in a switch debouncer. Capacitor C_1 smooths the pulses and the Schmitt trigger produces a single transition of the output voltage. This action depends on the hysteresis of the Schmitt trigger which is about 2.2 V for the 4093 at 10 V supply; for the 40106, hysteresis is guaranteed to be greater than 20% of the supply voltage.

Figure 9.4 shows the use of a Schmitt trigger as an astable. Only two passive

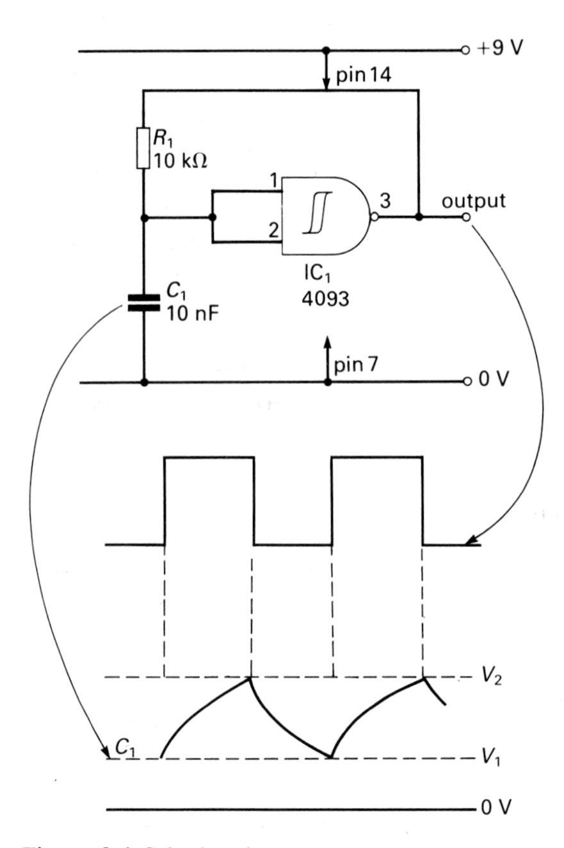

Figure 9.4 Schmitt trigger

components are needed, R_1 and C_1. The capacitor C_1 charges and discharges between the upper and lower threshold voltages causing the output voltage to produce a square wave. The frequency, f, of the square wave is given by $f = 1/(R_1C_1)$, approximately. Thus for $R_1 = 10\ \text{k}\Omega$ and $C_1 = 100\ \text{nF}$, $f = 1/(10^4 \times 10^{-7}) = 10^3 = 1\ \text{kHz}$, approximately.

9.4 Experiment E8

Looking at Schmitt trigger action

Figure 9.5 shows how to measure the hysteresis of a Schmitt trigger, e.g. the 40106. Set the variable resistor, VR_1, so that the output voltage is high (LED_1 on). Now slowly cover LDR_1 and note the reading on the voltmeter when LED_1 goes off (upper threshold voltage). Slowly uncover LDR_1 and note the reading of the voltmeter when LED_1 goes on again (lower threshold voltage). The difference between these two readings is the hysteresis of the Schmitt trigger. Note that this hysteresis is 'built into' digital Schmitt triggers; an op amp (analogue device) is needed (Book D, Chapter 4) if you want to adjust the amount of hysteresis.

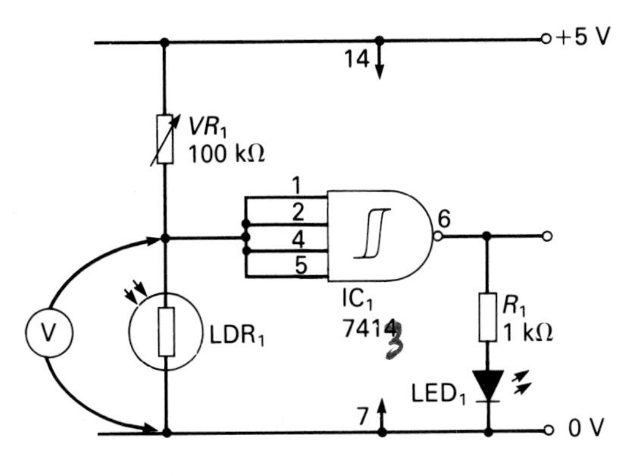

Figure 9.5 Experiment E8: Measuring the hysteresis of a Schmitt trigger

Figure 9.6 shows how to use a dual-beam oscilloscope to see how the Schmitt trigger works as an astable. Adjust the trigger control on the oscilloscope so that the input and output signals are displayed one above the other. Use the voltage sensitivity setting on the oscilloscope to measure the two threshold voltages. Use the time/division setting to work out the frequency of the astable and compare it with that obtained using $f = 1/(R_1C_1)$. (See Book A, Chapter 10, if you need help with using an oscilloscope.)

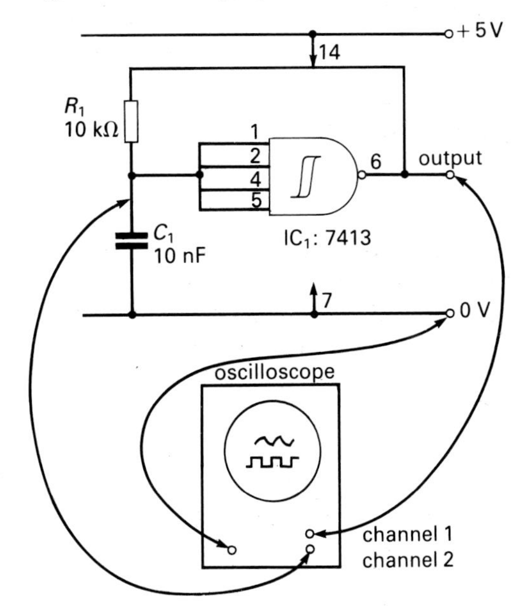

Figure 9.6 Looking at Schmitt trigger action

9.5 Bistable, astable and monostable multivibrators

Multivibrators are a family of two-stage transistor switching circuits in which the output of each stage is fed back to the input of the other stage using coupling capacitors or resistors. This feedback has the effect of driving the transistors alternately into saturation and cut-off so that while one output is high the other is low.

There are three types of multivibrator: the bistable, the astable and the monostable.

The *bistable*, or flip-flop (see Chapters 5 and 13), has two stable states. One state remains high and the other low until an external trigger switches the high state low and the low state high.

The *monostable* has one stable state and one unstable state; it can be switched to a high unstable state for a period of time determined by the time constant of a *CR* combination. The 555 timer (Book B, Chapter 14) is an example of a purpose-designed integrated circuit for use as a monostable.

The *astable* has no stable state; each of its two outputs switches from high to low automatically at a frequency determined by the time constant of a *CR* combination. The two-transistor astable is described in Book C, Chapter 25, and the use of the 555 timer as an astable is described in Book B, Chapter 14.

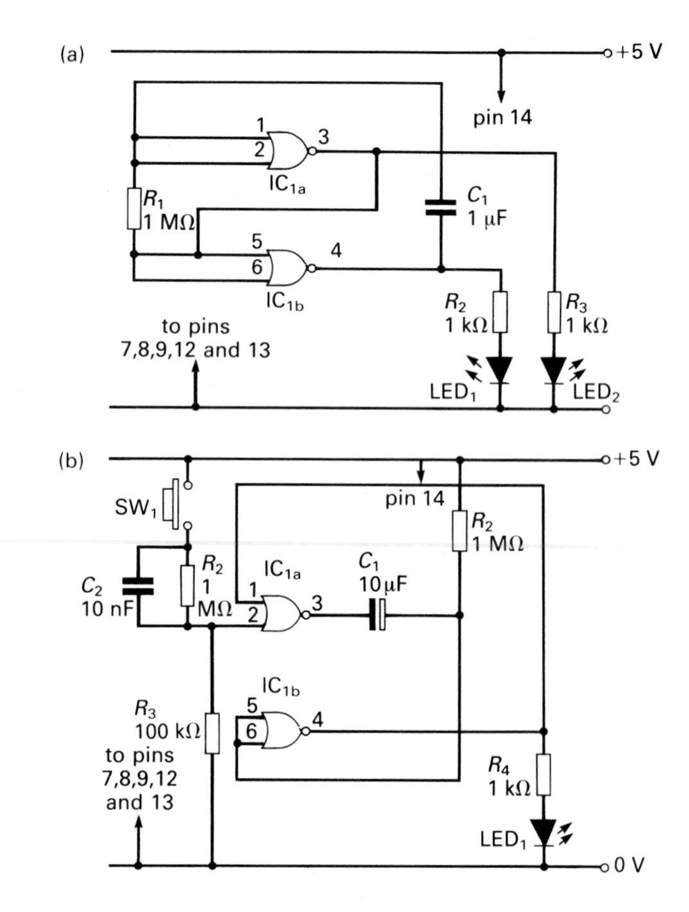

Figure 9.7 Astable and monostable using CMOS NOR gates: (a) astable using 4001 NOR gates; (b) monostable using 4001 NOR gates

9.6 Astables and monostables using NOR gates

In a circuit design, it is sometimes convenient to make use of spare logic gates to provide oscillations (astable) or a time delay (monostable). Figure 9.7 shows how you can use inexpensive CMOS gates to build an astable and a monostable from two NOR gates. Similar circuits can be built using NAND gates. TTL logic gates can be used in similar ways.

The function of each circuit shown in figure 9.7 is determined by the timing components R_1 and C_1. The frequency, f, of the astable is given by $f = 1/(R_1C_1)$, and the time delay, T, provided by the monostable, by $T = R_1C_1$. Note that in the astable C_1 must be an unpolarised (e.g. polyester) type of capacitor. In the monostable, C_1 may be an electrolytic type.

Thus the frequency of the square wave produced by the astable is equal to $1/(R_1C_1) = 1/(10^6 \times 10^{-6}) = 1$ Hz, approximately. The time delay of the monostable is equal to $R_1 \times C_1 = 10^6 \times 10^{-5} = 10$ s, approximately. The value of the timing resistor, R_1, in the astable can range from 4.7 kΩ to 10 MΩ, and the value of the timing capacitor from a few tens of picofarads to a few microfarads (limited by the fact that C_1 must be an unpolarised type). In the monostable, R_1 can range from 4.7 kΩ to 10 MΩ and C_1 from about 100 picofarads to a few thousand microfarads. Note that if a variable resistor is used in place of R_1, a fixed-value 2.2 kΩ resistor should be wired in series with it.

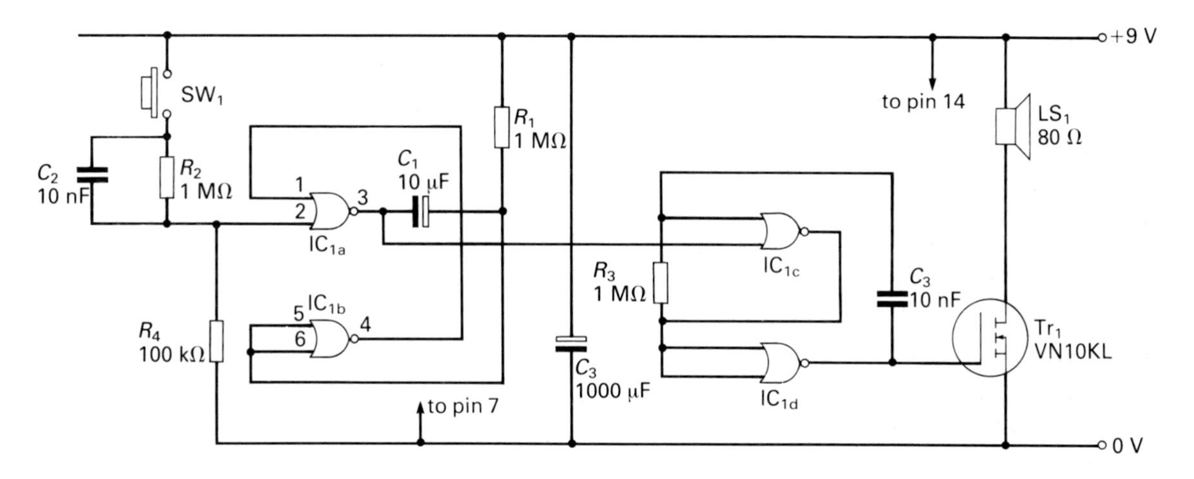

Figure 9.8 Auto-turn-off alarm call generator

Figure 9.8 shows how to wire up four NOR gates in a 4001 package to provide an auto-turn-off alarm call generator. Gates IC_{1a} and IC_{1b} are wired as a monostable, which turns on IC_{1c} and IC_{1d} wired as an astable. A tone of about 1 kHz is produced for about ten seconds. The pulse that starts the monostable timing is produced by momentarily pressing SW_1. Capacitor C_2 and resistor R_2 'debounce' SW_1. The VMOS transistor, Tr_1, (see Book C, Section 26.10) amplifies the audio tone which is heard from the loudspeaker, LS_1.

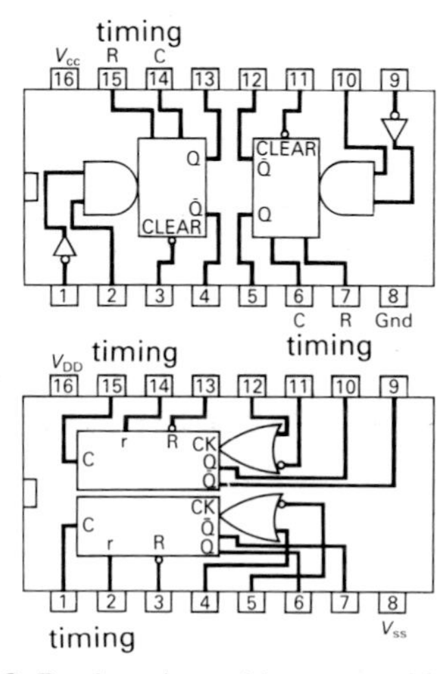

Figure 9.9 Dual retriggerable monostables

9.7 Monostable and astable IC packages

Ready-made monostables and astables are available in both the TTL and CMOS range of ICs. Figure 9.9 shows the TTL 74123 and the CMOS 4528 packages which are dual-retriggerable monostables. This means that the two monostables in each package can be used independently. The timing capacitor and resistor for each monostable has to be connected externally as shown in figure 9.10 for the 4528. Once triggered, the time delay can be restarted by pressing SW_1, i.e. the monostable is retriggerable.

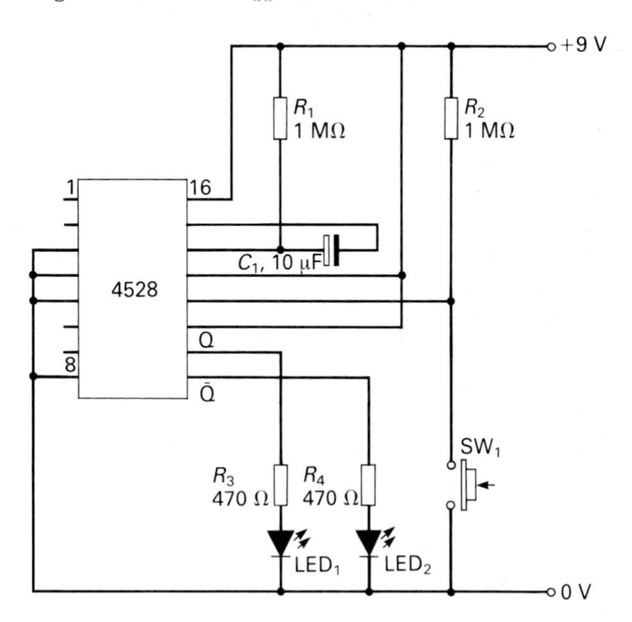

Figure 9.10 A test circuit for the 4528 monostables

Each monostable in the 4528 has complementary outputs, Q and $\overline{Q}$. Two trigger inputs are provided, pin 12 (or pin 4) enabling the monostable to be triggered from a rising voltage, and pin 11 (or pin 5) from a falling voltage. When SW_1 is pressed, a falling trigger voltage is applied to pin 11. Resistor R_2 normally holds pin 11 high.

Before SW_1 is pressed, the Q output will be low and the $\overline{Q}$ high. After SW_1 is released, the Q output will remain high (LED_1 lit) for time T given by $T = R_1C_1 =$ 10 s for the values shown. If SW_1 is pressed again after a time delay has been started, the time delay is restarted. Thus T begins after the last triggering edge arrives at the trigger input. Monostables like the 4528 and the 74123 are often used in circuits to lengthen a short duration pulse so that counters, for example, operate more reliably. In these applications, the monostables are called *pulse stretchers*.

A 4528 is used as a pulse stretcher in the design of the ultrasonic remote control described in Book D, Section 11.4. In the CMOS range, the 4047 device can be wired to operate as a monostable or as an astable depending on the connections of external resistors and capacitors. This makes the 4047 as versatile as the ubiquitous 555 timer.

The normal mode (there are alternatives) for operating the 4047 as an astable is shown in figure 9.11(a). The frequency obtained from the complementary outputs Q and $\overline{Q}$ is given by $1/(4.4R_1C_1)$ seconds, approximately, where R_1 is in MΩ and C_1 in μF. The range of permitted values for R_1 is 10 kΩ to a few megohms, and for C_1 from 100 pF to as high a value as possible for a non-polarised capacitor.

Figure 9.11(b) shows one version of several possible monostable configurations for the 4047. This circuit is a negative-triggered monostable, i.e. its output at Q and $\overline{Q}$ is set for a time delay when the input goes from logic 1 to logic 0. The Q output is normally low and produces a positive output pulse, and the $\overline{Q}$ output is normally high and goes low for the duration of the time delay. The approximate time delay is given by $2.48R_1C_1$ seconds where R_1 should be between 10 kΩ and a few megohms, and C_1 any value from 1000 pF upwards limited by the fact that it should be a non-polarised device.

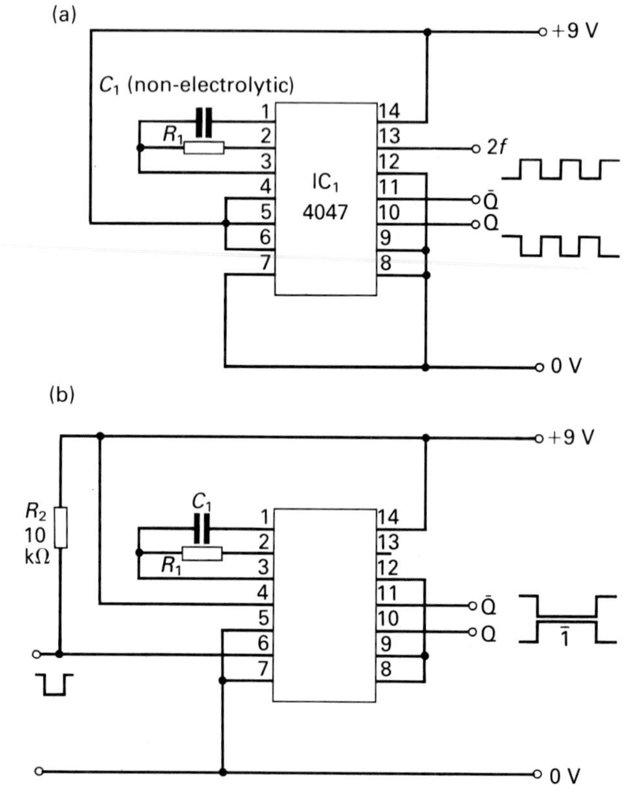

Figure 9.11 Using the 4047: (a) the normal way of operating it as an astable; $f = 1/(4.4\ R_1C_1)$ approx. (b) connecting it as a negative edge-triggered monostable; $T = 2.48\ R_1C_1$ seconds (approx.)

9.8 Crystal oscillators

Oscillators using transistors, logic gates, or purpose designed integrated circuits such as the 555 timer, do not have very stable output frequencies. When a stable and precise frequency is required, e.g. for use in clocks and watches, crystal oscillators are used.

Certain crystalline materials, in particular quartz, have a property called *piezoelectricity*. This means that if a crystal is squeezed, it produces a potential difference across some of its crystal faces. Conversely, if a potential difference is applied across these faces, the crystal responds by changing shape. A quartz crystal for use in crystal oscillators is a thin plate of quartz (or ceramic material) which has a natural resonant frequency. It is used in an oscillator circuit which stimulates this resonant frequency. The output frequency from a crystal oscillator is stable and has a very low temperature coefficient, i.e. the frequency changes very little with temperature change.

Quartz crystals are available with resonant frequencies ranging from 20 MHz down to 32.768 kHz. A 4 MHz crystal might well be used in a microcomputer for controlling the sequence of operations of its microprocessor. The 'odd' frequency of the 32.768 kHz crystal is chosen because, by continually dividing by 2 using a binary divider, it is possible to obtain a precise frequency of 1 Hz for use in clocks and watches. Thus if 32.768 kHz is divided by 2 fifteen times, a 1 Hz signal is obtained (see Section 6.2) Project Module D5, Frequency Divider, described in Chapter 13 of Book D, shows a circuit for producing these electronic 'ticks'.

Figure 9.12 shows the block diagram of a digital watch in which a quartz crystal stabilises the frequency of the on-chip oscillator to exactly 32.768 kHz. Extremely stable pulses from this oscillator are fed to a chain of frequency dividers which outputs a 1 Hz pulse. After counting and decoding by logic circuits in the decoders (see Chapter 7), signals are sent to the electrodes of the LCD segments so that time and date are displayed. The silicon chip in the watch of figure 9.13 contains all the major electronic functions of the watch.

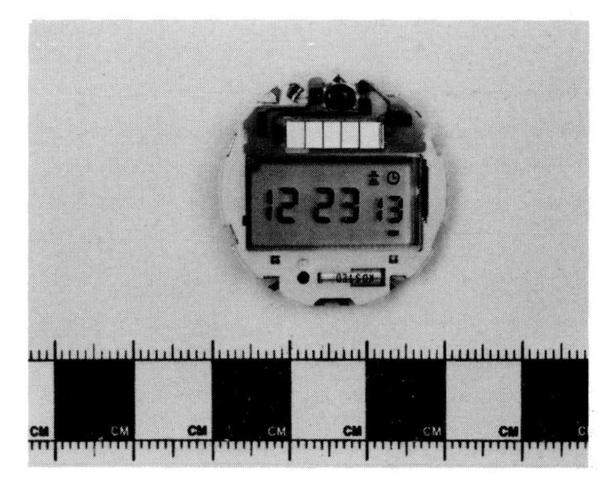

Figure 9.13 The inside of a digital watch – the bottle-shaped object is the crystal

Question

1 How many times does 4.194304 MHz have to be divided by 2 to obtain an output frequency of 1 Hz?

Quartz crystals are usually packaged in metal cans with two leads as shown in figure 9.14. Figure 9.15 shows how astable circuits using gates can accommodate a quartz crystal to produce stable square wave output frequencies.

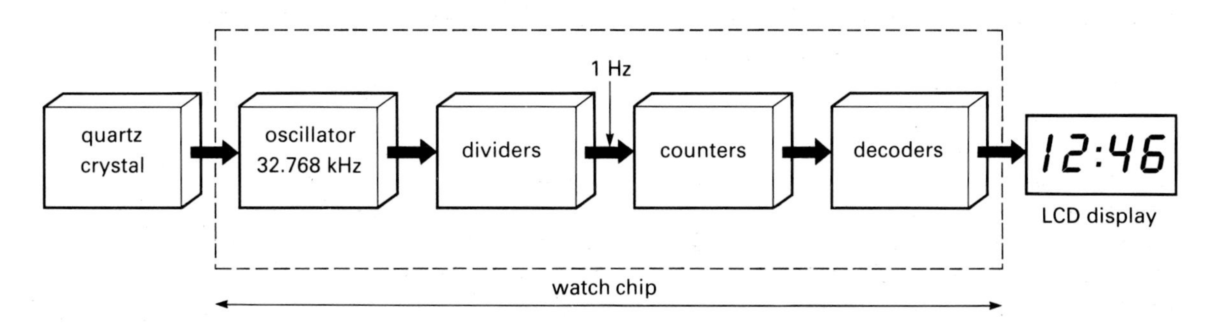

Figure 9.12 The building blocks of a digital watch

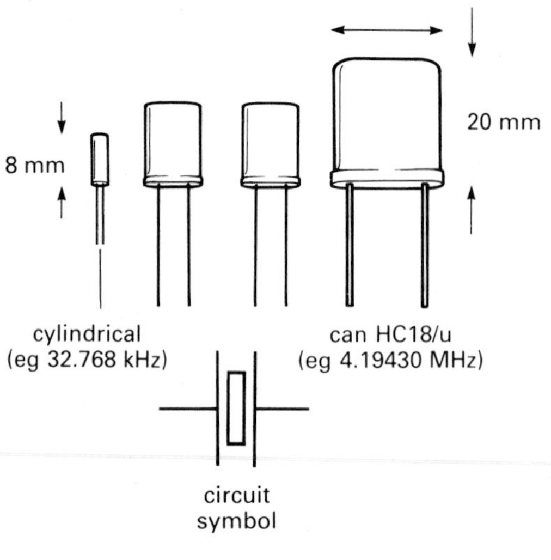

8 mm

20 mm

cylindrical
(eg 32.768 kHz)

can HC18/u
(eg 4.19430 MHz)

circuit
symbol

Figure 9.14 Types of packaging for quartz crystals

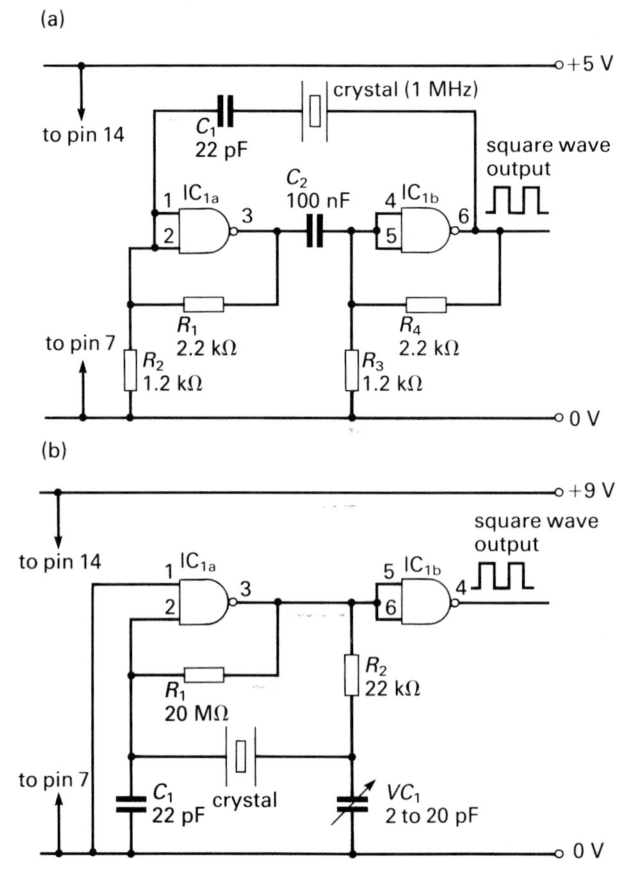

Figure 9.15 (a) TTL crystal oscillator using TTL NAND gates; (b) CMOS crystal oscillators using 4001 NAND gates

10 Number Systems

10.1 Decimal, binary and hexadecimal

When we count things in ordinary everyday activities, we use the *decimal number system*. This number system uses the ten familiar digits 0 to 9 to express the magnitude of any number we like.

Thus, the decimal number 273.16, two hundred and seventy three point one six, is constructed from the ten digits by giving a 'weight' to each digit. The weight depends on how far the digit is from the decimal point. Digits to the left of the decimal point increase their weights by progressively larger powers of ten while digits to the right of the decimal point have their weights reduced progressively by powers of ten.

The number 273.16 is constructed as shown in figure 10.1.

This decimal number is better expressed using the subscript '10', i.e. 273.16_{10}, to distinguish it from other number systems.

Question

1 Express the number 3621.5_{10} as a sum of powers of ten.

The decimal system is said to have a *radix*, or *base*, of ten. It is probably because we have ten fingers that we use a number system with a base of ten. However, it is said that the indigenous people of one of the islands in the Torres Straits in the Far East use just two digits in their number system called 'urapan' (0) and 'okasa' (1), even though they have the usual complement of ten fingers! They count using what we call the binary system which is explained in this chapter.

Along with the growth of digital electronics in recent years has come the more general use of the *binary system* and the *hexadecimal system*. The binary system uses two digits, and therefore has a base of two. The hexadecimal system has a base of sixteen.

For the binary system, the two decimal digits 0 and 1 are used. In the hexadecimal system not enough decimal digits are available for all the required sixteen digits. So the first ten digits use the decimal digits, 0 to 9, followed by the six letters A to F, to make up sixteen digits.
The first twenty numbers in the decimal, binary and hexadecimal number systems are shown in figure 10.2. The method of converting between these two number

Figure 10.1

	digit	weight	digit & weight	decimal number
	2	10^2	2×10^2	200
increasing	7	10^1	7×10^1	70
powers of	3	10^0	3×10^0	3
ten	1	10^{-1}	1×10^{-1}	.1
	6	10^{-2}	6×10^{-2}	.06
			sum:	273.16

systems is described in Sections 10.2 and 10.3.

Figure 10.2

decimal	hexadecimal	binary
0	0	0
1	1	1
2	2	10
3	3	11
4	4	100
5	5	101
6	6	110
7	7	111
8	8	1000
9	9	1001
10	A	1010
11	B	1011
12	C	1100
13	D	1101
14	E	1110
15	F	1111
16	11	10000
17	12	10001
18	13	10010
19	14	10011
20	15	10100

A binary digit, 0 or 1, is known as a *bit* which is probably a contraction of *bi*nary digi*t*, or possibly it just means a small piece of information. In computing jargon, four bits, e.g. 1101_2, make a *nibble* and 8 bits make a *byte*. The general name for a bit pattern made up of nibbles and bytes is a *word* — some people call it a *gulp*!

A hexadecimal number is commonly used to simplify a binary word of 8-bit bytes into easily remembered two-character numbers. For example, the decimal number, 119_{10} has a binary equivalent of 01110111_2. In hexadecimal this binary byte is simply 77_{16}. Note the use of the subscript '2' and '16' to identify a binary and hexadecimal number, respectively. It is possible to find a few shortcuts for converting numbers between the binary, decimal and hexadecimal number systems.

10.2 Converting binary to decimal

First note the general principle for expressing a number according to its base. We saw, above, how the number 273.16_{10} was constructed. This is,

$$273.16_{10} = 2 \times 10^2 + 7 \times 10^1 + 3 \times 10^0 + 1 \times 10^{-1} + 6 \times 10^{-2}$$

Now in binary we are dealing with powers of 2, not 10. For example,

$$\begin{aligned} 1101.01_2 &= 1 \times 2^3 + 1 \times 2^2 + 0 \times 2^1 \\ &\quad + 1 \times 2^0 + 0 \times 2^{-1} \\ &\quad + 1 \times 2^{-2} \\ &= 8 + 4 + 0 + 1 + 0 + 0.25 \\ &= 13.25_{10} \end{aligned}$$

Thus it is only necessary to convert the powers of two into their decimal values and then to add the products. A faster way you might prefer to use is to list the value of the base to different powers over each bit. Then add these values over the '1' bits and ignore those over the '0' bits. Thus

$$\begin{array}{llllll} 32 & 16 & 8 & 4 & 2 & 1 \\ 1 & 1 & 0 & 1 & 0 & 1_2 = 32 + 16 + 4 + 1 = 53_{10} \end{array}$$

Question

1 Convert 10011.1_2 into decimal.

10.3 Converting decimal to binary

A short way of doing this conversion is to repeatedly divide the decimal number by two and to record the remainder after each division. These remainders are either 0 or 1 and form the binary number. For example, suppose the number 202_{10} is to be converted to binary.

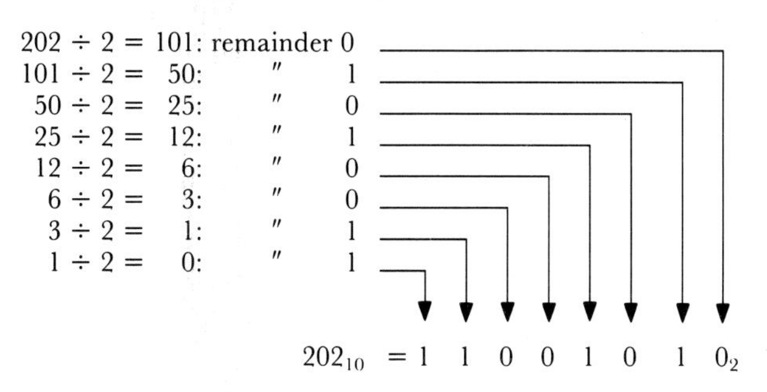

If decimal fractions are to be converted, e.g. 0.3_{10}, the rule is to *multiply* by two for digits on the right of the decimal point and to record the whole number integer in the forward direction. Thus

$0.3 \times 2 = 0.6$: integer = 0
$0.6 \times 2 = 1.2$: " 1
$0.2 \times 2 = 0.4$: " 0
$0.4 \times 2 = 0.8$: " 0
$0.8 \times 2 = 1.6$: " 1
etc.

$0.3_{10} = 0.0\ 1\ 0\ 0\ 1_2$

The multiplication is carried on until you obtain the required accuracy in the conversion.

Questions

1 Convert 67_{10} into binary.
2 Convert 99.9_{10} into binary.

10.4 Converting hexadecimal to decimal

Since hexadecimal numbers have a base of sixteen, the decimal number should be expressed as a sum of powers of sixteen. Thus $F7_{16}$ becomes

$$\begin{aligned} F7_{16} &= F \times 16^1 + 7 \times 16^0 \\ &= 15 \times 16^1 + 7 \times 16^0 \\ &= 247_{10} \end{aligned}$$

Similarly,

$$\begin{aligned} C2E_{16} &= C \times 16^2 + 2 \times 16^1 + E \times 16^0 \\ &= 12 \times 256 + 2 \times 16 + 14 \times 1 \\ &= 3118_{10} \end{aligned}$$

Question

1 Convert $A3_{16}$ into decimal.

10.5 Converting decimal to hexadecimal

All that is required is to divide the decimal number by sixteen, repeatedly. For example,

$59_{10} \div 16 = 3$, remainder 11 (B_{16})
$3_{10} \div 16 = 0$, remainder 3

$59_{10} = 3\ B_{16}$

If the decimal number is fractional, the digits to the right of the decimal point should be *multiplied* by sixteen and the result expressed as for the decimal to binary conversion. Thus

$0.7_{10} \times 16 = 11.2$: integer = 11 (= B_{16})
$0.2 \times 16 = 3.2$: " = 3
etc.

$0.7_{10} = 0.B\ 3_{16}$

Questions

1 Convert 82_{10} into hexadecimal.
2 Convert 23.8_{10} into hexadecimal.

10.6 Converting binary to hexadecimal

If the binary number is a byte long, first divide the byte into two nibbles and then write down the hex equivalents of the nibbles. Thus,

$$1110\ 0110_2 = 1110\ 0110 = E6_{16}$$

If the binary number is not a byte long, the following examples will make the conversion clear.

$$0111_2 = 7_{16}$$
$$111011_2 = 0011\ 1011_2 = 3B_{16}$$
$$1101101101_2 = 0011\ 0110\ 1101_2 = 3AD_{16}$$

Questions

1 Convert 1110_2 into hexadecimal.
2 Convert 10001111001_2 into hexadecimal.

10.7 Converting hexadecimal to binary

This conversion is simply a matter of assigning the binary equivalent to each hex digit. Thus

$$F8_{16} = 1111\ 1000_2$$
$$7D4 = 011111010100_2$$

Questions

1 Convert $D3_{16}$ into binary.
2 Convert FF_{16} into binary.

10.8 Binary arithmetic

First let's remind ourselves of how to add two decimal numbers, e.g. 239_{10} and 823_{10}.

D	C	B	A
	2	3	9
	8	2	3
1	0	6	2_{10}

Starting with column A, the addition of 9 and 3 gives 12. This is greater than the base 10 of the decimal system so 2 is recorded in the 'units' column A and 1 is carried over to the 'tens' column B. Next the contents of column B are added giving a total of 6 (tens) which is less than the base 10 so there is no carry to column C. This procedure is repeated for column C.

If the same rules are applied to the addition of two binary numbers, e.g. 11010_2 and 10011_2, the addition is carried out as follows.

	E	D	C	B	A	binary addition rules
	1	1	0	1	0	$0 + 0 = 0$
	1	0	0	1	1	$0 + 1 = 1$
						$1 + 0 = 1$
1	0	1	1	0	1	$1 + 1 = 1\ 0$ (0 carry 1)
						$1 + 1 + 1 = 10 + 1 = 11$

In column A, $0 + 1 = 1$ which is less than the base 2. So there is no carry digit. Next for column B, $1 + 1 = 10_2$ since $1_2 + 1_2$ is equal to the base 2 and a 0 is recorded in the 'twos' column and a carry digit goes to the 'two squared' column C. Likewise the addition is carried to the remaining columns.

Question

1 Add 1010_2 to 1011_2 and express your answer in binary and decimal.

In subtracting two decimal numbers, e.g. taking 187_{10} away from 359_{10}, the following procedure is used:

D	C	B	A
	3	5	9
	1	8	7
	1	7	2_{10}

Starting with the right hand column, the subtraction of 7 from 9 leaves 2. But in column B, 8 cannot be taken from 5 unless a carry digit is borrowed from the 'hundreds' column C, which reduces the digit 3 to 2 and makes the 5 in column B, 15. Subtracting 8 from 15 gives 7 which is recorded in the difference column. Then, for column C, 1 taken away from 2 (not 3) gives a difference of 1 in the 'hundreds' column C.

Similar rules are obeyed when doing subtraction in binary, e.g. subtracting 10010_2 from 11001_2.

E D C B A	binary subtraction rules
1 1 0 0 1	0 − 0 = 0
1 0 0 1 0	1 − 0 = 1
	1 − 1 = 0
0 0 1 1 1	0 − 1 = 1 borrow 1
	(or = − 1)
	10 − 1 = 1

In column A, 1 − 0 = 1. In column 2, 0 − 1 = 1 since 2 (the base) is borrowed from column C. This is put back in column 3 and again 0 − 1 = 1 after borrowing from column 4.

Questions

2 Subtract 10101_2 from 11010_2.

3 Subtract 11111_2 from 10000_2.

10.9 Hexadecimal arithmetic

Many computers use binary numbers which are processed in groups of four or eight bits each. With these computers, it is convenient to use the hexadecimal equivalent of the binary numbers. For example, $C6_{16}$ and $F8_{16}$ are added together as shown below. In column P, 6 + 8 = E_{16} which is less than the base sixteen so an E is recorded in the units column A and there is no carry digit to the 'sixteens' column Q.

R Q P	decimal equivalent
C 6	$C \times 16^1 + 6 = 12 \times 16^1 + 6 = 198$
F 8	$F \times 16^1 + 8 = 15 \times 16^1 + 8 = 248$
1 B E	$1 \times 16^2 + B \times 16^1 + E$
	$= 256 + 176 + 14 = 446_{10}$

Next, for column Q, C + F = $1B_{16}$ since the addition is greater than the base sixteen and a carry digit is moved to the 'sixteen squared' column R.

Question

1 Add AB_{16} to $2F_{16}$ and express your answer in hexadecimal and decimal.

Of course, the hexadecimal quantities can be converted to decimal and the decimal addition used to check your answer as shown in the example.

The subtraction of hexadecimal numbers is just as easy provided you remember that the base of hex is sixteen. The following two examples show the technique.

(a) R Q P	(b) Q P
E 6	B 5
8 2	A D
6 4_{16}	0 8_{16}

In column P of (a), 6 − 2 = 4_{16} as in decimal since we are subtracting equal digits in both systems. In column Q, E − 8 = 6 as you can see by writing out the sixteen hexadecimal digits.

In the second example (b), 5 − D requires a borrow of '16', the base of hexadecimal numbers, from the top digit in column Q. Thus, 15 − D = 8_{16} and the B in column Q is reduced to A. And for column Q, A − A = 0.

Questions

2 Subtract $D8_{16}$ from $F9_{16}$. Express your answer in hexadecimal and decimal.

3 Subtract $2B_{16}$ from $B9_{16}$. Express your answer in hexadecimal and binary.

11 Binary Codes

▽ 11.1 The 8421 code

Any decimal number which is expressed in binary digits as a series of 1s and 0s is called a *natural binary code*. Thus the decimal number 347_{10} has a natural binary code of 101011011_2.

Although calculators, computers and numerically controlled machines readily handle data in binary coded form, the size of a binary number such as 101011011_2 is not easily understood until it is converted into decimal. However, there is a way of presenting binary numbers which enables their magnitude to be more easily interpreted.

If the binary number contains more than 4 bits, it is divided into *decades*. A decade contains four bits representing any one of the decimal numbers from 0 to 9. To represent a large decimal number, the binary coded decimal (BCD) code is used – see figure 11.1.

Figure 11.1

hundreds decade	tens decade	units decade
3	4	7
0011	0100	0111

Thus the BCD number for 347_{10} is 001101000111 which is nothing like the natural binary code for 347_{10}. If we know a binary number is in BCD form, it is easy to estimate its decimal equivalent. For example, if the binary number 10010100111001 is known to be in BCD form, its decimal equivalent is found by dividing into four bit groups starting from the right and assigning a digit of 0 to 9 as follows

0010	0101	0011	1001
2	5	3	9

i.e. the decimal equivalent is 2539_{10}.

Since the weightings of the four bits in any binary number are in the ratio 8:4:2:1 reading from left to right, the BCD code is called an *8421 code*.

As with all groups of binary digits, it is usual to give the letters A, B, C, and D to the four bits of a BCD number. Bit A is called the least significant bit (l.s.b.) and bit D is the most significant bit (m.s.b.). Figure 11.2 summarises the BCD code. Note that the codes from 1010 (10_{10}) to 1111 (15_{10}) are forbidden numbers in the BCD code. Thus the BCD code for the decimal number 12 is 0001 0010, not 1100 as for the binary equivalent of 12. The main advantage of the BCD code is the ease of conversion to and from binary and decimal numbers. The BCD code is a mixed-base code since it is binary within each four-bit group and decimal from group to group.

Figure 11.2

decimal number	binary coded decimal number			
	D (m.s.b.)	C	B	A (l.s.b.)
0	0	0	0	0
1	0	0	0	1
2	0	0	1	0
3	0	0	1	1
4	0	1	0	0
5	0	1	0	1
6	0	1	1	0
7	0	1	1	1
8	1	0	0	0
9	1	0	0	1

11.2 The Gray code

The Gray code is a code in which each Gray number differs from the preceding number by a single bit. Figure 11.3 compares the Gray and binary codes for the first 15 decimal numbers.

Figure 11.3

decimal number	binary number	Gray code
0	0000	0000
1	0001	0001
2	0010	0011
3	0011	0010
4	0100	0110
5	0101	0111
6	0110	0101
7	0111	0100
8	1000	1100
9	1001	1101
10	1010	1111
11	1011	1110
12	1100	1010
13	1101	1011
14	1110	1001
15	1111	1000

The Gray code is an unweighted code which is not suited to arithmetic operations. But it is useful in digital input and output devices, e.g. shaft encoders (figure 11.4), since only one bit changes from one position to the next which reduces the possibility of errors in angular measurement. For instance, in going from decimal 3 to decimal 4, the Gray code changes from 0010 to 0110 which differ in only one bit. In the binary code, a change from decimal 3 to 4 causes all three bits to change from 0011 to 0100. Thus the Gray code reduces 'bit chatter', i.e. oscillation between a 0 and 1 level at the point of transition. The encoding disc shown in figure 11.5 is designed for use in a wind direction indicator which provides a 4-bit code for each of the 16 points of the compass.

Figure 11.4 Shaft encoder

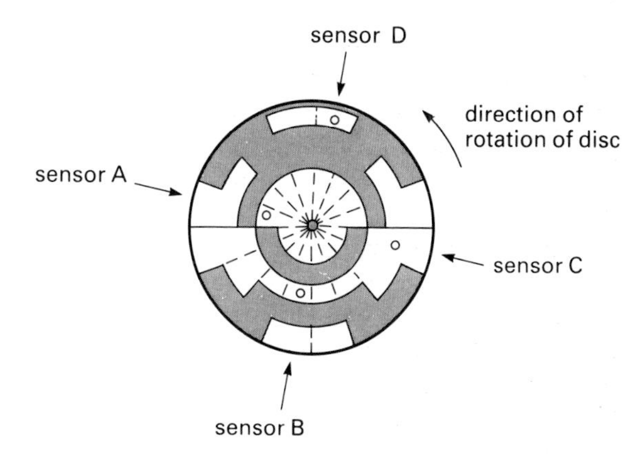

Figure 11.5 Encoding disc which uses the Gray code

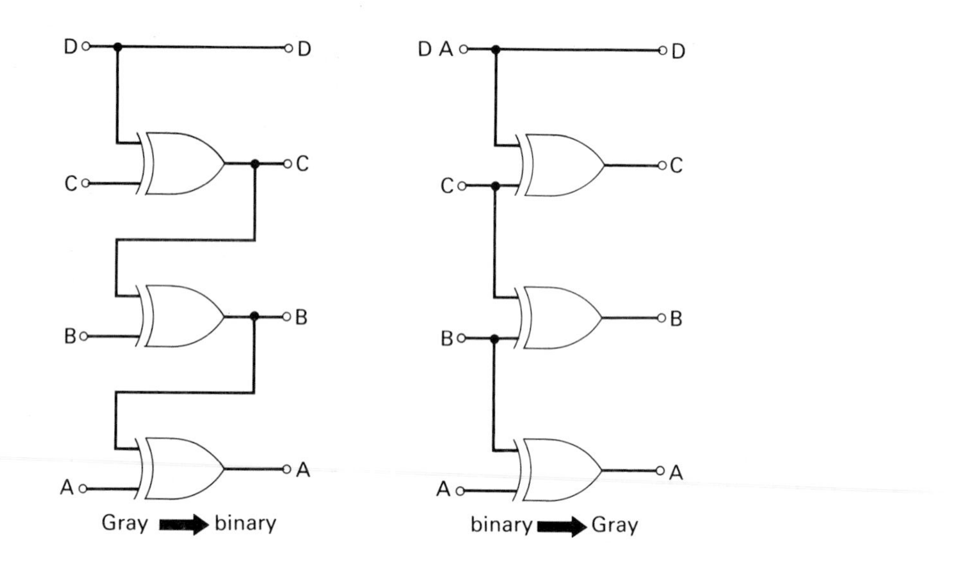

Figure 11.6 Using XOR gates to convert from (a) Gray to binary and (b) binary to Gray

The rules for converting Gray code to binary code are as follows:

For the conversion, start with the m.s.b. and work towards the l.s.b.

(a) Leave the m.s.b. unchanged.

(b) If the number of 1 bits to the left of the next Gray code bit is odd, change that bit. If it is even, leave unchanged.

(c) Repeat instruction (b) until conversion is complete. For example

Gray code	1 1 0 1
(a)	1 – – –
(b)	1 0 – –
(c)	1 0 0 –
(c)	1 0 0 1

Thus the binary equivalent is 1001_2. Here is another example.

Gray code	1 1 1 0
	1 – – –
	1 0 – –
	1 0 1 –
	1 0 1 1

Thus the binary equivalent is 1011_2.

The rules for converting binary code to Gray code are as follows:

Again start by converting the m.s.b. of the binary number and work towards the l.s.b.

(a) leave the most significant bit unchanged

(b) add each pair of adjacent bits starting with the m.s.b. to get the next Gray digit, and ignore any carries.

For example:

binary code	1 0 0 1
(a)	1 – – –
(b) $1 + 0 = 1$	1 1 – –
(b) $0 + 0 = 0$	1 1 0 –
(b) $0 + 1 = 1$	1 1 0 1

Thus the Gray code equivalent of 1001_2 is 1101.

Questions

1 Convert the following Gray codes into binary numbers:
(a) 0011
(b) 1110.

2 Convert the following binary codes into Gray codes:
(a) 1111
(b) 1010.

Though the Gray code may look complex, figure 11.6 shows how easy it is to convert between 4-bit binary and Gray codes using exclusive-OR logic gates. Note that the most significant bit (m.s.b.) is common to both 4-bit codes. △

▽ 11.3 The ASCII code

This American Standard Code for Information Interchange is widely used in computer systems. It is a set of 128 characters comprising letters, numbers, punctuation marks, and symbols, each represented by an 8-bit binary word. The ASCII code facilitates the exchange of information between a computer and other data processing equipment, e.g. keyboards and printers.

Most microcomputers use the ASCII code but often with slight variations to chosen graphics symbols and some omissions from the standard set of characters. The 7-bit word associated with each character is stored in the computer's *read-only memory* (ROM). Should the computer be asked to print the character whose code is 61 (0111101 in binary, 3D in hex), the result is the character =, i.e. the equals sign.

The ASCII code set used by the Sinclair Spectrum or the BBC Acorn microcomputers can be obtained by typing in:

```
10 FOR c=32 TO 255: PRINT CHR$ c;:
NEXT c
```

△

12 More Boolean Algebra

▽ 12.1 Basic Boolean expressions

As explained in Chapter 4, Boolean algebra simplifies writing down statements which are logical. Thus the logical statement 'Alice and Bill or Cheryl and Don will be in the show' can be written in shorthand as A.B + C.D = S (S = 'in the show'). Remember that the functions of the seven basic logic gates — AND, OR, NOT, NAND, NOR, XOR and XNOR — can be written in the shorthand of Boolean algebra as follows:

AND: $A.B = S$ OR: $A + B = S$
NOT: $\overline{A} = S$ NAND: $\overline{A.B} = S$
NOR: $\overline{A + B} = S$ XOR: $A \oplus B = S$
XNOR: $\overline{A \oplus B} = S$

Remember, too, that the bar over the top of a symbol indicates that the value of the symbol is complemented, i.e. inverted, so that $\overline{0} = 1$ and $\overline{1} = 0$. Since a symbol, e.g. A, can be either 0 or 1, the following simple Boolean expressions are true:

(i) $A + 1 = 1$ (ii) $A.1 = A$
(iii) $A + A = A$ (iv) $A.A = A$
(v) $A + \overline{A} = 1$ (vi) $A.\overline{A} = 0$

Thus, for expression (ii), if A = 1, 1.1 = 1; if A = 0, 0.0 = 0 which is true for an AND gate. Check the other expressions by letting A = 0 and A = 1.

It was also shown in Chapter 4 that 'universal' NAND gates can be combined together to simulate the function of these basic logic gates. The NOR and XOR functions using NAND gates were proved using truth tables. Let's see how two further Boolean expressions can be used to prove the NOR and XOR functions using NAND gates, and to reduce other complex logic circuits to combinations of NAND (or NOR) gates.

12.2 De Morgan's Theorems

De Morgan was a friend of George Boole and he produced two further useful Boolean expressions which are known as De Morgan's Theorems. They are:

$\overline{A + B} = \overline{A}.\overline{B}$ which can be written: 'the complement of two or more symbols ORed together is the ANDed result of the complement of each symbol'

$\overline{A.B} = \overline{A} + \overline{B}$ which can be written: 'the complement of two or more symbols ANDed together is the ORed result of the complement of each symbol'

These equations can be extended to include more symbols, e.g. $\overline{A + B + C} = \overline{A}.\overline{B}.\overline{C}$.

De Morgan's Theorems can easily be proved by writing down the truth table for the two sides of each expression (figure 12.1).

Figure 12.1

A	B	$\overline{A}$	$\overline{B}$	$\overline{A + B}$	$\overline{A}.\overline{B}$	$\overline{A.B}$	$\overline{A} + \overline{B}$
0	0	1	1	1	1	1	1
0	1	1	0	0	0	1	1
1	0	0	1	0	0	1	1
1	1	0	0	0	0	0	0
				1st theorem		2nd theorem	

De Morgan's first theorem is true since the 5th and 6th columns are identical. And De Morgan's second theorem is true since the 7th and 8th columns are identical.

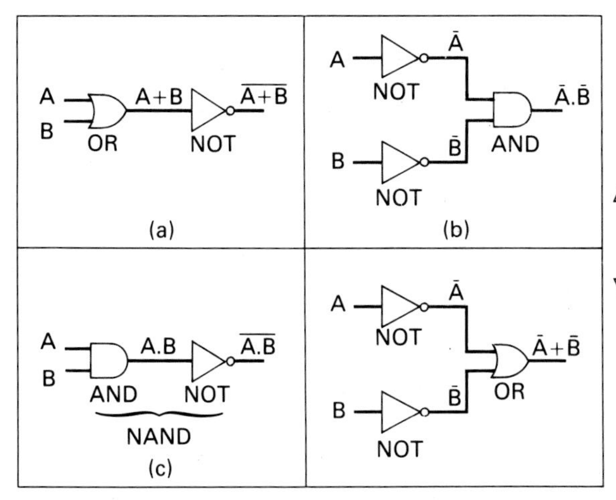

Figure 12.2 Logic gate equivalents of De Morgan's Theorem

Figure 12.2 shows the logic gate equivalents of De Morgan's Theorems. The truth table in figure 12.1 shows that the logic circuit of figure 12.2(a) is equivalent to that of figure 12.2(b). You can easily prove that the circuits are equivalent by converting the OR, AND and NOT gates to NAND gates using the circuits of figure 4.2.

Thus, figure 12.2(a) is a NOR gate. A NOR gate is made from four NAND gates as shown in figure 4.2(d), i.e. two NOT gates followed by an AND gate (a NAND gate followed by a NOT gate).

Figure 12.2(c) is a NAND gate. If the OR gate in figure 12.2(d) is converted to three NAND gates as shown in figure 4.2(c), i.e. two NOT gates followed by a NAND gate it produces a single NAND gate as required. △

▽ 12.3 Simplifying logic circuits

In addition to De Morgan's Theorems, and the expressions in Section 12.1, there are a few other Boolean expressions which are helpful in designing logic circuits. These are:

Commutative laws:

$$A + B = B + A$$
$$A.B = B.A$$

Associative laws:

$$A + (B + C) = (A + B) + C$$
$$A.(B.C) = (A.B).C$$

Distributive law:

$$A.(B + C) = A.B + A.C$$

Thus the function of the logic circuit of figure 12.3 is given by the Boolean expression S = A.C + A.B.C. The circuit

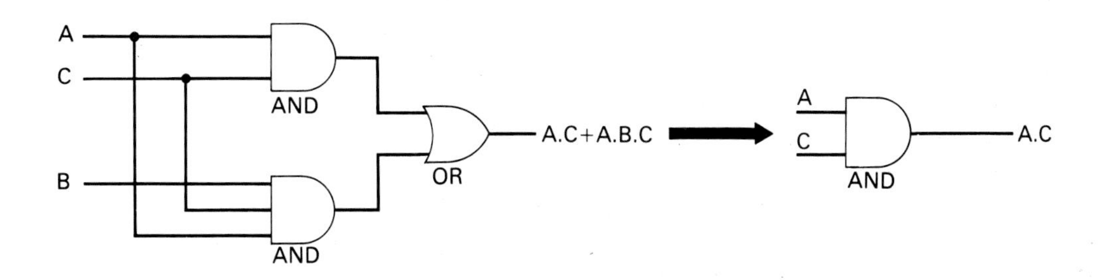

Figure 12.3 This logic circuit is equivalent to ANDing A and C

is actually equal to an AND gate since the expression reduces to A.C(1 + B) = A.C using the above Distributive law and the expression (1 + B) = 1 — see Section 12.1.

Example 1 Voting System

As a practical example of the use of Boolean algebra in logic design, suppose you are asked to design the following voting system.

Three voters and one chairman can cast votes. The chairman is asked to throw a casting vote if the three voters do not reach a unanimous decision. If the chairman votes 'yes' he adds 1 to the total vote; if he votes 'no' he subtracts 1 from the total vote. The vote is carried, i.e. the answer is 'yes', if 2 or 3 votes in total (including the chairman's vote) are cast in its favour, otherwise the vote is 'no'.

The truth table in figure 12.4 summarises all possible combinations of votes cast. The table is divided into two parts. The first part shows the outcome when the chairman votes 'no' and a 1 is subtracted from the total 'yes' votes. The second part shows the outcome when the chairman votes 'yes' and a 1 is added to the total 'yes' votes.

This truth table gives the following Boolean equation which states the possible combinations of votes from A, B, C and D that provide a 'yes' vote, i.e. a 1, in the last column. Thus:

$$Y = (A.B.C) + (A.D) + (B.D) + (A.B.D) + (C.D) + (A.C.D) + (B.C.D) + (A.B.C)$$

This equation can be simplified first by factorising just as in conventional algebra. Thus:

$$Y = (A.B.C) + (A.B.C) + C.D.(1 + A + B) + A.D + B.D(1 + A)$$

Now

$$(A.B.C) + (A.B.C) = (A.B.C)$$
(just as A + A = A)

Figure 12.4

chairman's vote D	**voters C**	**B**	**A**	**total 'yes' votes**	**outcome of vote**
not asked	0	0	0	0	0
0	0	0	1	0	0
0	0	1	0	0	0
0	0	1	1	1	0
0	1	0	0	0	0
0	1	0	1	1	0
0	1	1	0	1	0
not asked	1	1	1	3	1
not asked	0	0	0	0	0
1	0	0	1	2	1
1	0	1	0	2	1
1	0	1	1	3	1
1	1	0	0	2	1
1	1	0	1	3	1
1	1	1	0	3	1
not asked	1	1	1	3	1

And

$$(1 + A + B) = 1, \text{ and } 1 + A = 1.$$

Thus

$$Y = (A.B.C) + (C.D) + (A.D) + (B.D)$$

or

$$Y = (A.B.C) + D.(A + B + C)$$

This is the simplest Boolean expression of the function of the voting system, and considerably simpler than the one we started with.

It is easy to convert this voting function into a logic circuit as shown in figure 12.5.

This circuit uses a 3-input AND gate, a 3-input OR gate, a 2-input AND gate and a 2-input OR gate. The logic circuit can be rationalised to use NAND and NOR gates as shown in figure 12.6.

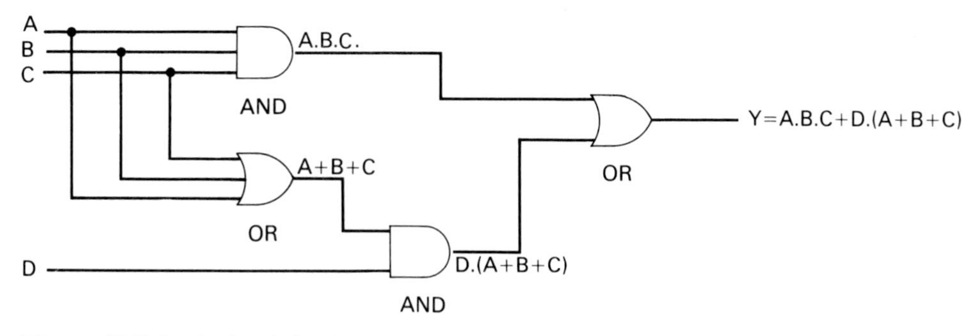

Figure 12.5 Logic circuit for the voting system

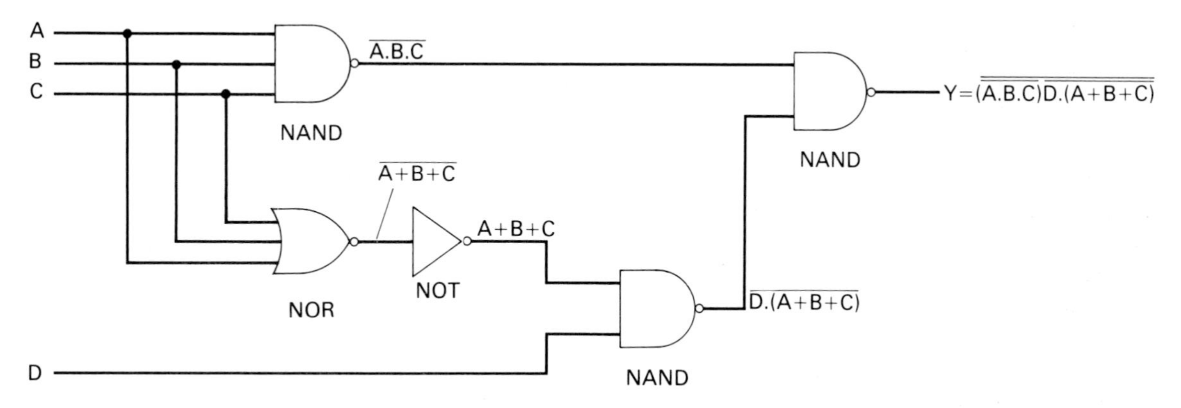

Figure 12.6 Logic circuit for the voting system using NAND and NOR gates

This solution follows by rewriting the equation for the voting system. Thus:

$$Y = \overline{\overline{(A.B.C) + D.(A + B + C)}}$$

This double inversion leaves the original equation unaltered. Now use the first of De Morgan's Theorems in Section 12.2. Thus:

$$Y = \overline{\overline{(A.B.C)}.\overline{D.(A + B + C)}}$$

This is the output from the logic circuit of figure 12.6 which uses NAND and NOR gates. Using CMOS logic, the 3-input NOR gate could be from the 4025 package; the two 2-input NAND gates could be obtained from two of the four 2-input NAND gates in the 4023 package; the other two 2-input NAND gates could provide the 3-input NAND gate function; and the inverter from one of the 3-input NOR gates in the 4025 package. Thus the circuit could be made using just two IC packages.

Example 2 Animal house

An animal is kept in a cage which has two sets of double doors as shown in figure 12.7. If both doors A and B or doors C and D are open at the same time, a warning siren sounds. Write down the Boolean expression for the conditions that the siren sounds. Then devise a logic circuit using NAND gates which could provide the warning required.

The Boolean expression A.B + C.D describes the action required. This expression can be converted to a combination of NAND statements using the first of De Morgan's Theorems as follows:
Apply a double complement to the equation:

$$\overline{\overline{A.B + C.D}}$$

which leaves the expression unchanged.

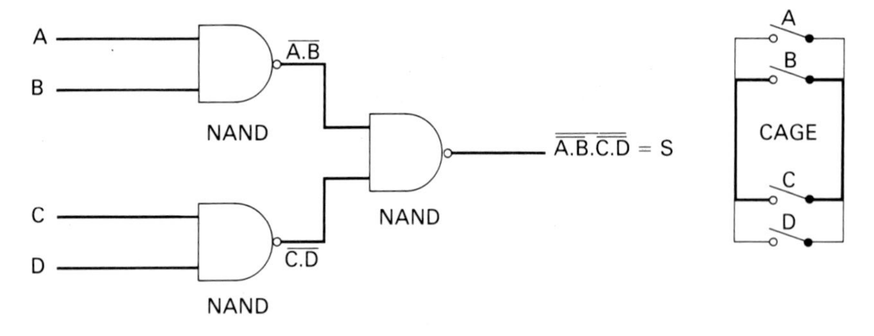

Figure 12.7 Logic circuit for the animal house problem

Apply De Morgan's Theorem for changing OR functions to AND functions:

$$\overline{\overline{A.B} + \overline{C.D}}$$

which is two NAND functions NANDed together.

The logic implementation of this function is shown in figure 12.7. You may check that this gives the alarm required, by drawing up a truth table in which 'door open' = 1, 'door closed' = 0, and 'alarm sounds' = 1.

The logic circuit could be achieved using a single 7400 or 4011 quad 2-input NAND gate package.

Questions

1 A room which has a window and a door is to be protected with an alarm system. The alarm system is armed with a keyswitch, K. The alarm sounds if either the window, W, or the door, D, is opened. Design a logic circuit which uses NOR gates to solve the problem.

2 In order that a bank safe can be opened, both the manager, M, and the cashier, C, or the cashier and the director, D, have to be present. Describe this arrangement using NAND gates.

3 Three grain bins are being filled automatically. Produce a logic circuit using 2-input NAND gates which provides an alarm when two of the bins are filled. △

13 More About Flip-flops

13.1 Reminder

Chapter 5 explained that a flip-flop:

(a) has two stable output states and can be triggered into one of these states by a suitable input pulse;

(b) is a memory unit since it stores data after the input pulse has passed;

(c) is a sequential logic unit since its output state depends on previously stored data as well as on any new data;

(d) can be made to 'toggle', i.e. act as a divide-by-two frequency divider.

Chapter 5 described the use of JK flip-flops in integrated circuit packages as binary counters (e.g. the TTL 7493) and as frequency dividers (e.g. the CMOS 4040). As well as the JK flip-flop, the RS flip-flop and the D-type flip-flop are available in IC packages. The design of these three flip-flops is explained in the following Sections.

Questions

1 What is the maximum factor by which a 4-bit counter can divide an input frequency?
2 Pulses at a frequency of 100 Hz are input to a 'toggle' flip-flop. What is the output frequency?
3 What is a BCD counter?

13.2 The RS flip-flop

The basic building block of all flip-flops in IC packages comprises two cross-coupled NAND gates, G_1 and G_2, as shown in figure 13.1. The output of each gate is connected to the input of the other. The RS flip-flop has two inputs, RESET (R), and SET (S); and two outputs Q and $\overline{Q}$. In the design shown in figure 13.1, the R and S inputs are applied to the cross-coupled gates via two NAND gates, G_3 and G_4, connected as inverters. Remember that a logic 0 on any input to a NAND gate gives a 1 output unless all inputs are at logic 0.

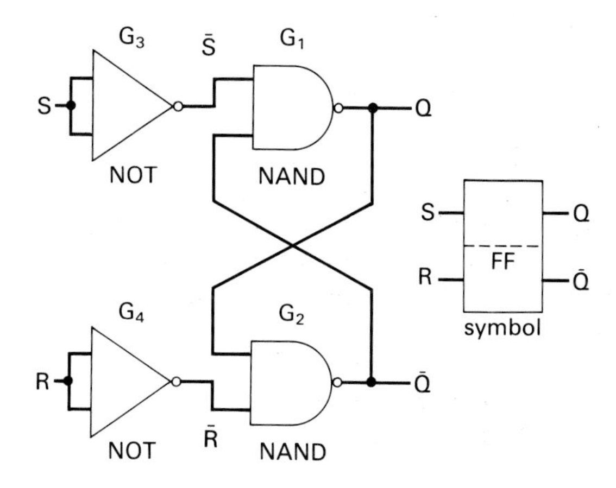

Figure 13.1 Basic RS flip-flop

The truth table in figure 13.2 shows what happens to the Q and $\overline{Q}$ outputs of this flip-flop when pulses are applied to the R and S inputs.

Figure 13.2

S	R	$\overline{S}$	$\overline{R}$	Q	$\overline{Q}$	action
1	0	0	1	1	0	set
0	0	1	1	1	0	no change
0	1	1	0	0	1	reset
0	0	1	1	0	1	no change
1	1	0	0	1	1	Stable
0	0					indeterminate

The first line shows the condition when S = 1 and R = 0. Then Q = 1 and $\overline{Q}$ = 0 so the flip-flop is said to be 'set'.

The second line shows R = 0 and S = 0. This means that Q and $\overline{Q}$ can be either 1 or 0 depending on the previous state of the output, i.e. there is no change in the output state.

The third line shows R = 1 and S = 0. Then Q = 0 and $\overline{Q}$ = 1 so the flip-flop is said to be 'reset'.

The fourth line shows R = 0 and S = 0. As for line 2, there is no change in the output state.

The input condition R = S = 1 makes both Q = 1 and $\overline{Q}$ = 1, which is unlike the other three cases. This is is a stable condition. However, if S and R change to 0 simultaneously, it is not possible to predict the states of the Q and $\overline{Q}$ outputs. A *race condition* occurs since one of the gates (G_1 or G_2) will switch faster than the other and it is not possible to predict whether Q or $\overline{Q}$ will be 1. The circuit action is said to be *indeterminate* and the condition S = R = 1 should not arise.

By modifying the input circuit to the RS flip-flop as shown in figure 13.3, a clocked RS flip-flop is produced. Before the clock pulse rises from 0 to 1, the outputs from the inverters, G_2 and G_3, are both logic 1.

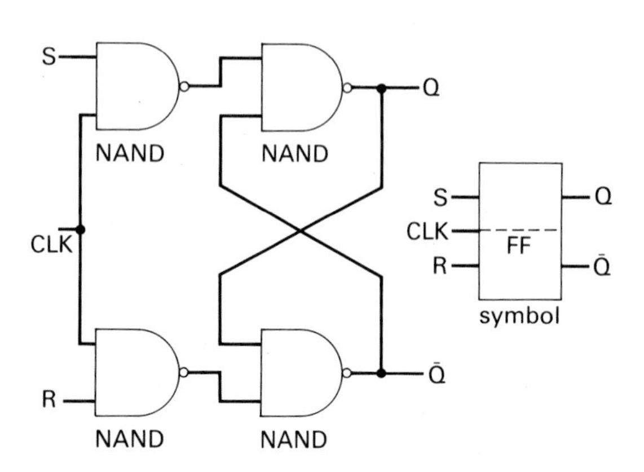

Figure 13.3 Clocked RS flip-flop

This corresponds to the 2nd and 4th rows in the truth table above, and so the Q and $\overline{Q}$ outputs retain their present values, i.e. 1 or 0 or 0 and 1, irrespective of changes in the values of S and R. The flip-flop is said to be *disabled* or *inhibited.*

When the clock pulse rises from 0 to 1, the flip-flop is enabled and the Q and $\overline{Q}$ outputs depend on the values of the S and R inputs. Thus the RS flip-flop is *level triggered* since changes to the outputs only occur while the clock is high. Moreover, the circuit does not remove the ambiguity in the truth table when R = S = 1. Improved designs use *edge triggering* where the input information can only be transferred to change the state of the flip-flop when the clock pulse is rising or falling. This is the basis of operation of D-type and JK flip-flops.

13.3 The D flip-flop

A D flip-flop can be made from an RS flip-flop by adding an inverter to the RESET input as shown in figure 13.4. Like an RS flip-flop, a D flip-flop has a data input, D, a clock input, CLK, and two complementary outputs, Q and $\overline{Q}$.

The D flip-flop is often called a *delay flip-flop* since the data, i.e. a 0 or a 1, at the input can't get to the output, Q, until the next clock pulse goes high. This behaviour is shown in the truth table in figure 13.5.

Note that as the clock pulse goes from 0 to 1, the bit of data on the D input gets passed to the Q output. At the end of the clock pulse, i.e. after the pulse level has returned from 1 to 0, the data on Q is retained. The Q output remains at 1 or 0 while the CLK input remains at 0, and does not change even if D changes.

The D flip-flop is called a *data latch* since the Q output holds on to the bit of data and stores it temporarily. D flip-flops are used in shift registers — see Section 13.5.

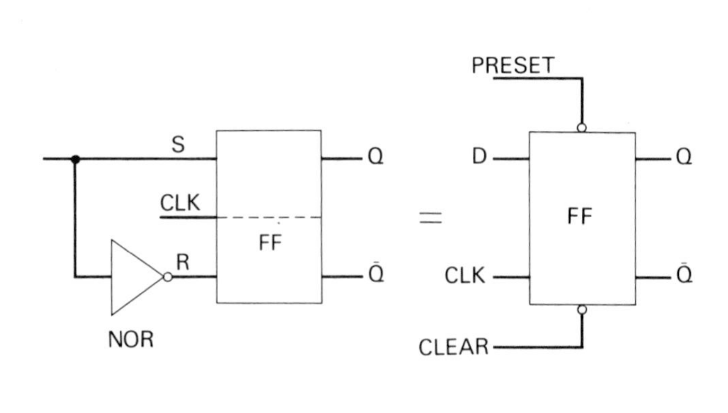

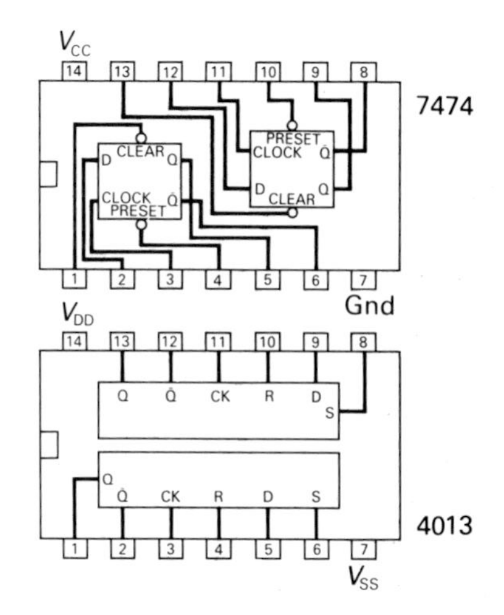

Figure 13.4 Symbol and packages of D flip-flops

Figure 13.5

input			**outputs before clock pulse**		**outputs after clock pulse**	
D	**S**	**R**	**Q**	**$\overline{Q}$**	**Q**	**$\overline{Q}$**
0	0	1	1	0	0	1
0	0	1	0	1	0	1
1	1	0	1	0	1	0
1	1	0	0	1	1	0

Note that the D flip-flop can be made to toggle, i.e. the Q and $\overline{Q}$ outputs change on successive clock pulses, if the $\overline{Q}$ output is connected to the D input. See if you can work out the sequence of changes to the output as the clock pulses change.

Figure 13.4 shows that the TTL 7474 and the CMOS 4013 devices are dual D flip-flops in 14-pin packages. Note that these flip-flops are edge triggered, i.e. the change in the output states takes place on the rising or falling part of the clock waveform. Note, also that they have PRESET and CLEAR inputs. If these inputs are high, the D flip-flop will toggle on the low to high edge of the clock pulses.

13.4 The JK flip-flop

The TTL 7476 and 4027 CMOS devices contain two independent JK flip-flops. They are used in Chapter 5 to show how flip-flops are the basis of binary counters. In these applications, the JK flip-flops were made to toggle, i.e. on repeated clock pulses the Q and $\overline{Q}$ outputs repeatedly turn on and off. This toggling action is the basis of divide-by-two counters. Four-bit binary counters (Chapter 5) and frequency dividers (Chapter 6) are produced by cascading flip-flops as in the 7493 and 4516 4-bit counters.

The connections on a commercial JK flip-flop are shown in figure 13.6: it has two data inputs, J and K; a clock, CLK, input; a PRESET, PS, input and a CLEAR, CLR, input; and the usual complementary outputs, Q and $\overline{Q}$. These connections give the JK flip-flop all the features of other types of flip-flops and it is widely used in circuit design. Most modern flip-flops have designs based on the master–slave principle as shown in figure 13.6. This design cures a problem which sometimes arises in digital counting called the *race hazard*. This problem can develop if the output of a flip-flop goes high while the clock input is still high. If this high output signal 'races around' to the input (due to the necessary feedback round the circuit) while the clock pulse is still high, overlap of the clock and feedback pulses occurs which makes flip-flops unreliable. Designers of digital circuits are always on their guard against racing!

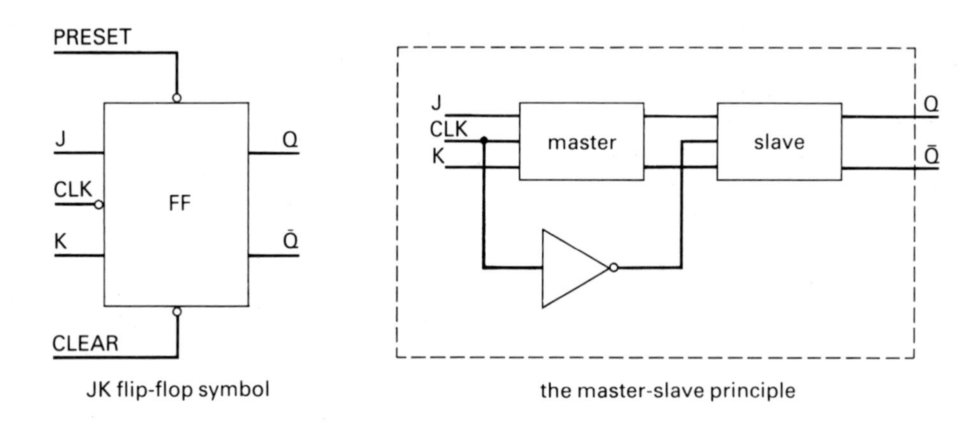

Figure 13.6 Symbol and master–slave principle of JK flip-flop

The master–slave design ensures that the data is only transferred to the output after the clock pulse has returned to low. The operation of the master–slave JK flip-flop occurs in two stages: in the first stage, during the rising clock pulse, the master is enabled allowing the data on the JK inputs to be stored. During this time, the slave is disabled by the inverter at the clock input. In the second stage, during the falling edge of the clock pulse, the master is disabled and the slave enabled so that the latter accepts data from the master and passes it to the output.

The truth table for the 7476 flip-flop is shown in figure 13.7. It shows the Q or $\overline{Q}$ output does one of four things: stays the same, resets, sets or toggles.

Figure 13.7

inputs				outputs after falling edge of clock pulse		
PS	**CLR**	**J**	**K**	**Q**	**$\overline{Q}$**	**Q or $\overline{Q}$ status**
H	H	L	L	?	?	stays the same
H	H	L	H	L	H	resets
H	H	H	L	H	L	sets
H	H	H	H	H/L	L/H	toggles
L	H	X	X	H	L	sets
H	L	X	X	L	H	resets

Line 1: Here both the J and K inputs are low, and the outputs have the values 0 or 1, just before the falling edge of the clock pulse, i.e. the outputs stay the same.
Line 2: The K input is high and the J low. The outputs are reset, i.e. Q goes low and $\overline{Q}$ high.
Line 3: The J input is high and the K input low. The outputs are set, i.e. Q goes high and $\overline{Q}$ low.
Line 4: Both the J and K inputs are high. The Q and $\overline{Q}$ outputs toggle, i.e. switch high and low on each falling edge of the clock pulse.
Line 5 and 6: Regardless of the states of the J and K inputs and the clock, a low on the PRESET input sets the Q output high and the $\overline{Q}$ low. A low on the CLEAR input sets the Q output low and the $\overline{Q}$ output high.

13.5 Shift registers

A shift register is an interconnected group of flip-flops that momentarily stores a binary number, and shifts it out as required. One flip-flop is used for each bit of stored data. Shift registers are important building blocks in digital systems. They are used in arithmetical circuits for the multiplication and division of binary numbers. They are also the basis of ring counters.

You can see shift registers at work when you use a pocket calculator. When you enter number 7, say, from the keyboard, it appears on the extreme right of the calculator display. Since it stays there when the key is released, it must have been stored. When you press key 2, say, number 7 is shifted one place to the left and 2 appears on the extreme right of the display. This example shows the memory and shifting characteristics of a shift register.

There are two methods of shifting binary numbers into a register: one bit at a time as in a serial shift register; or all bits at the same time as in a parallel shift register. Figure 13.8 shows a 4-bit shift register using D flip-flops (e.g. the TTL 7474 or CMOS 4013 devices — see Section 13.3). The following tables show how the data, input one bit at a time via the data input of FF1, appears as a parallel data word on the four outputs, QA to QD. The CLEAR input enab!es all these outputs to be set to 0 simultaneously.

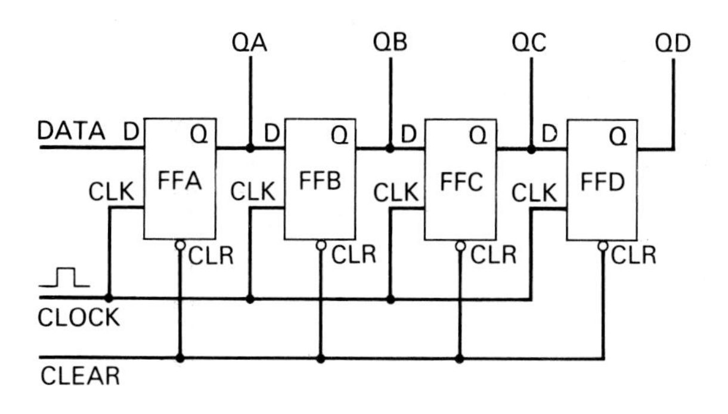

Figure 13.8 A 4-bit serial load shift register using D flip-flops

Suppose the content of the register is as follows:

QA	QB	QC	QD
1	0	0	1

The content of the register after the next clock pulse is:

QA	QB	QC	QD
0	1	0	0

i.e. the content have been shifted right by one place and a 0 has been shifted into the end of the register from which a 1 has been shifted.
The content of the register after the next clock pulse is:

QA	QB	QC	QD
0	0	1	0

The shifting of the data through the flip-flops of a shift register allows us to take data from the shift register one bit at a time, or to feed bits into a register and assemble them one bit at a time. This is the basis by which signals are sent over telephone lines, radio links or to printers attached to computers.

Shift registers are also used in microprocessors to carry out multiplication of binary numbers. This is because a left shift is equivalent in binary numbers to multiplication by 2, just as a left shift of decimal numbers is equivalent to multiplication by 10. Thus the decimal number 169 becomes 1690 (ten times larger) when the digits are shifted one place to the left. Similarly, the binary number 0011 (decimal 3) becomes 0110 (two times larger) when the digits are shifted one place to the left. This is just the way the pocket calculator works; the first digit entered is in units but it becomes the tens digit when the second digit entered shifts the first digit one place to the left.

Two types of shift register in IC packages are shown in figure 13.9. The 4015 CMOS device contains two identical 4-bit shift registers and the test circuit

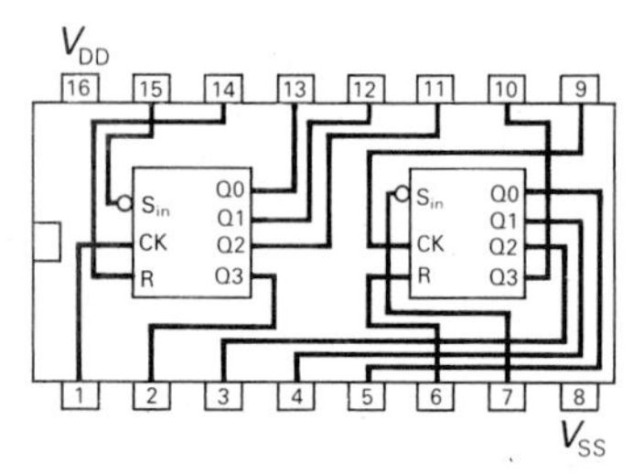

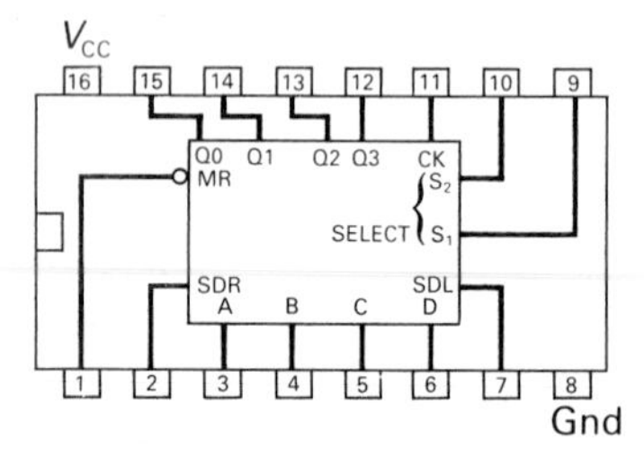

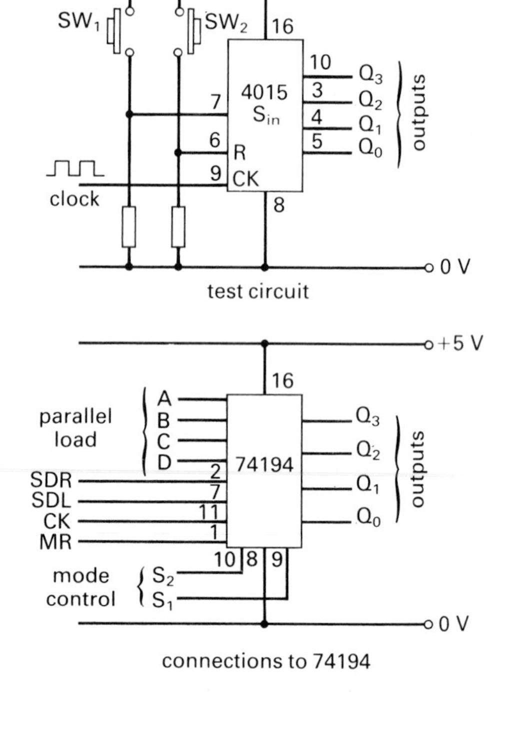

Figure 13.9

shows how to see one of them in action. Switch SW_2 is pressed to bring pin 6 high and reset the internal flip-flops to zero so that the output reads 0000. When switch SW_1 is pressed, a 1 bit passes into the shift register on the next high-to-low change of the clock pulse: if it is left open, a 0 bit passes into the shift register. The 1 or (0 bit) appears at pin 5 so that the output would read 0001. This bit is shifted through the register on each successive clock pulse. You could use LEDs at the outputs to show where the bits are in the register.

The 74194 TTL device is more versatile than the 4015. This is a 4-bit bi-directional universal shift register. It is bi-directional because bits can be shifted to the right or left; and it is universal since it can be used as a serial or parallel load shift register. Two or more 74194s can be cascaded to make an 8-bit or longer shift register. These various uses are selected by the mode control terminals, S_1 and S_2, which can have the codes shown in figure 13.10.

Figure 13.10

shift action	S_1 and S_2 code
parallel load	$S_1 = 1, S_2 = 1$
shift right	$S_1 = 1, S_2 = 0$
shift left	$S_1 = 0, S_2 = 1$
inhibit	$S_1 = 0, S_2 = 0$

Thus with $S_1 = S_2 = 1$, and the MR (CLEAR) input (pin 1) high, data on the A, B, C and D inputs appears on the outputs on the low-to-high transition of the clock pulse.

13.6 *Experiment* E9

Building a bidirectional shift register

You can explore what the 74194 bidirectional shift register can do by assembling the circuit shown in figure 13.11a on breadboard – figure 13.11b shows how. The circuit using a binary-coded switch, SW_1, which has four binary-weighted output terminals on which the 4-bit word appears, and a common terminal which is connected to V+. This switch provides a 4-bit word and is operated by a screwdriver to 'load' a 4-bit binary word into the parallel inputs of the shift register, IC_2. Note you need a 5 V supply for this 74194 TTL circuit.

The operation of the shift register is regulated by the clock, IC_1, which is based on a 555 timer wired as an astable. It produces clock pulses at about one second intervals which is monitored by LED_1. Four more LEDs monitor the value of the binary word available at the Q outputs of the shift register. They enable you to see where the 4-bit data is and how it can be shifted left or right through the register.

There are three flying leads, R, L and X, so that you can get data to load, four bits at a time, shift it right and shift it left. Figure 13.9 shows that leads R and L are connected to the two mode terminals S_1 and S_2 of the register. Lead X is connected to 0 V. When the circuit is up and running and clock pulses are being delivered by IC_1, this is what you do with these leads.

(a) First let the three leads X, R and L be 'free-flying' – don't connect them anywhere, even to each other. This allows the S_1 and S_2 control inputs to float high.

(b) Rotate SW_1 to select different 4-bit words and you will see each word appear on LEDs 2 to 5 with each low-to-high change of the clock pulse, ie as LED1 lights. Thus with both S_1 and S_2 high, parallel data is loaded into the register at each clock pulse.

(c) For a particular setting of SW_1 giving, say, a binary number of 0110 (decimal 6), touch together leads X and R. This makes S_1 = 0 and S_2 = 1 and is the condition for data to be shifted right. You should see the following sequence on the LEDs.

clock	Q_D	Q_C	Q_B	Q_A	
L	0	1	1	0	
H	1	0	1	1	Data moved one place right
L	1	0	1	1	Data stays put
H	1	1	0	1	Data moved next space right
L	1	1	0	1	Data stays put
H	1	1	1	0	Data moves next space right
L	1	1	1	0	And so on
H	1	1	1	1	

Now disconnect leads X and R and allow a clock pulse to reload to 0110 into the register. Touch leads X and L together, ie S_2 = 0 and S_1 = 1, and note that the binary word is shifted to the left on each low-to-high change of the clock pulse.

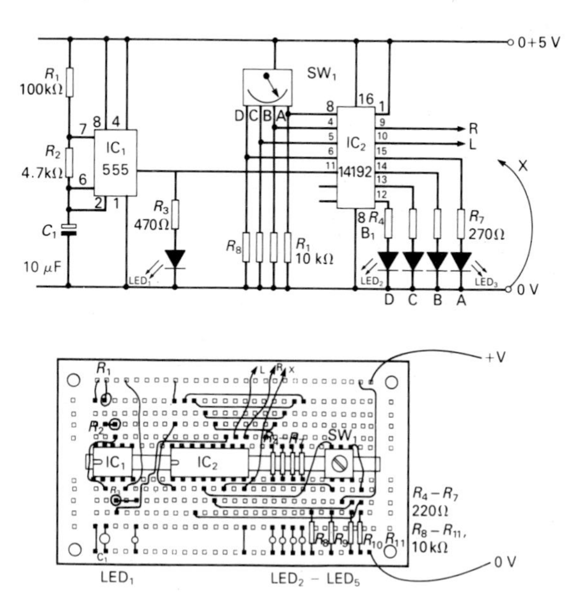

Figure 13.11

14 Arithmetic Circuits

14.1 Introduction

Though Blaise Pascal's father was a mathematician, he forbade his son access to any books on mathematics. Instead he wanted him to study ancient languages. But Blaise so impressed his father with his understanding of geometry that his father gave in and let the boy study mathematics. In 1642 when he was nineteen, Blaise Pascal invented a calculating machine that, by means of cogged wheels, could add and subtract. This machine was the ancestor of the modern mechanical cash register which has now given way to the electronic machines that perform arithmetic at lightning speed.

For example, microcomputers and calculators manipulate binary numbers using circuits that add and subtract. In a microcomputer these circuits are to be found in the microprocessor. This is the 'brain' of not only microcomputers but also washing machines, computer printers, video players, robots and even toasters! The key part of the microprocessor is the *arithmetic and logic unit (ALU)*. It contains the electronic logic circuits that add and subtract and which perform logical operations such as ANDing and ORing. The adding of binary numbers is done with logic gates as shown below.

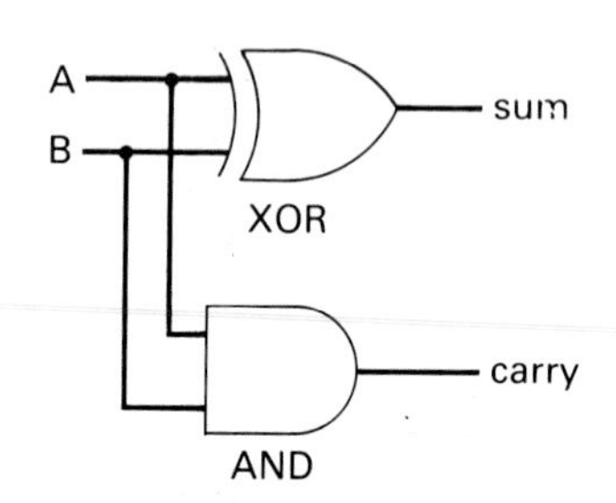

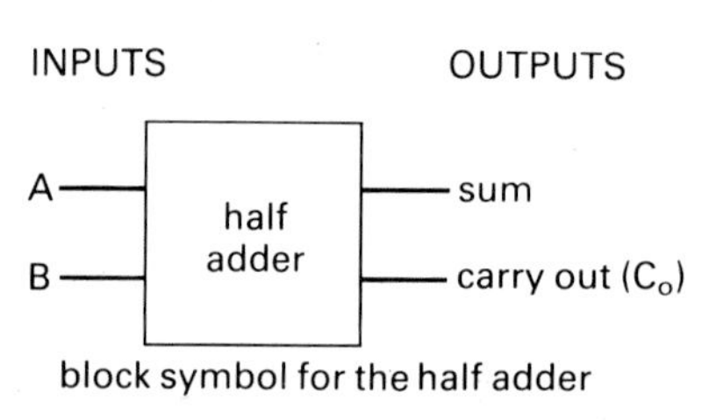

Figure 14.1 Logic diagram for the half-adder and its symbol

14.2 The half-adder

The half-adder does what we do mentally when we add two binary digits (see Section 10.8). As figure 14.1 shows, a half-adder can be made from an exclusive-OR gate and an AND gate. Since the circuit has two inputs, there are four (2^2) combinations of inputs to consider and they are summarised in figure 14.2.

Figure 14.2

A	B	carry = A.B	sum = A $\oplus$ B
0	0	0	0
0	1	0	1
1	0	0	1
1	1	1	0
bits to be added		AND gate	XOR gate

The last line shows why an XOR gate is required since 1 + 1 = 0 and carry 1, i.e. 0 is the sum and 1 is the carry. Thus the half-adder can only deal with the addition of the least significant bit. Clearly, we need a system which not only adds A and B but copes with the carry bit. Such a system is called a *full-adder.*

14.3 The full-adder

A circuit that can add three binary digits is known as full-adder. For instance:

$$\begin{array}{r} 111 \\ +\ 101 \\ \hline 1100 \end{array}$$

In the right-most column, 1 + 1 = 0 and carry 1. In the next column, three digits must be added because of the carry, i.e. 1 + 0 + 1 = 0 and carry 1. In the third column, three digits must again be added, i.e. 1 + 1 + 1 = 1 and carry 1.

The full-adder which can do this addition of three binary digits is shown in figure 14.3. The truth table in figure 14.4 summarises all the additions the full-adder can carry out.

Two examples of the full-adder at work are shown in figure 14.5. For instance, suppose A = 1, B = 1 and C = 0 as shown in figure 4.5(a). The first half-adder has a sum of 0 and a carry of 1. The second half-adder has a sum of 0 and a carry of 0. Thus the final output is a sum of 0 and a carry of 1 as summarised in line 4 of the truth table. If A = 1, B = 1 and C = 1 as shown in figure 14.5(b), line 8 of the truth table indicates that we get a sum of 1 and a carry of 1.

Figure 14.4 Using full-adders

inputs			outputs	
carry in, C_{in}	**B**	**A**	**sum**	**carry out, C_o**
0	0	0	0	0
0	0	1	1	0
0	1	0	1	0
0	1	1	0	1
1	0	0	1	0
1	0	1	0	1
1	1	0	0	1
1	1	1	1	1

carry + B + A

(a)

HA HA OR

A = 1, B = 1, C_{in} = 0
output = 0, carry 1

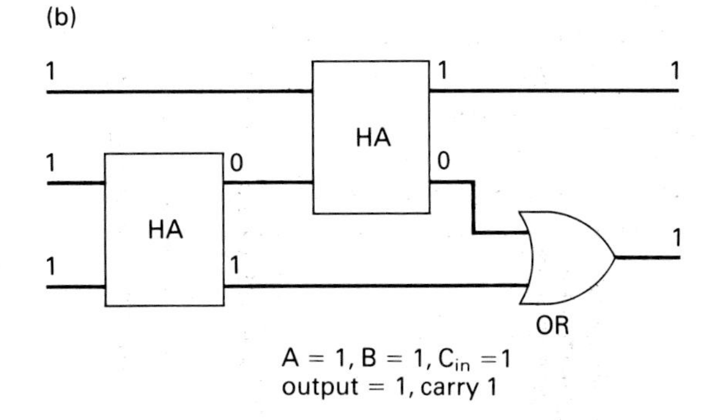

Figure 14.5 Examples of full-adders in operation.

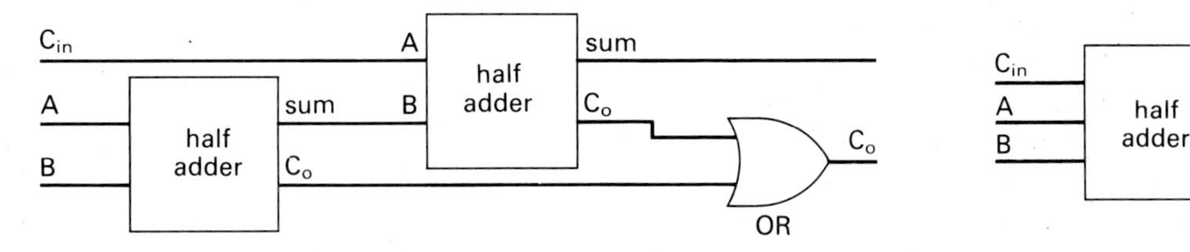

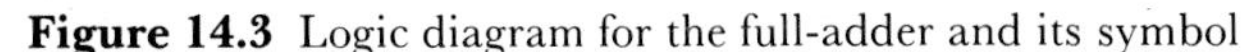

Figure 14.3 Logic diagram for the full-adder and its symbol

14.4 4-bit IC adders

It is easy to add two 4-bit numbers using integrated circuits designed for the job. In the TTL range of digital ICs, the 7483 is a 4-bit binary full adder, and a similar device in the CMOS range is the 4008. These two packages are shown in figure 14.6.

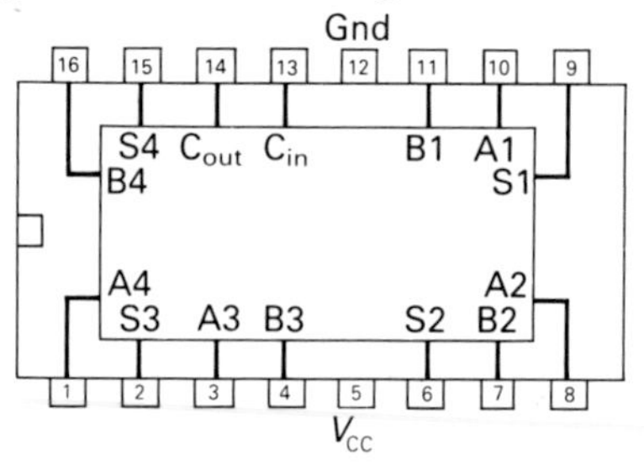

TTL 4-bit full-adder (7485)

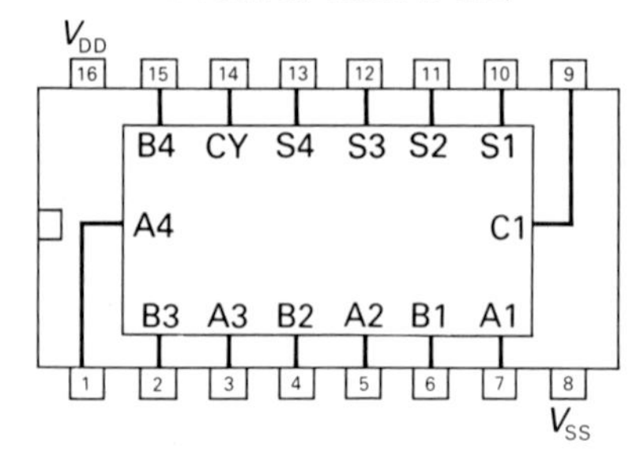

CMOS 4-bit adder(4008)

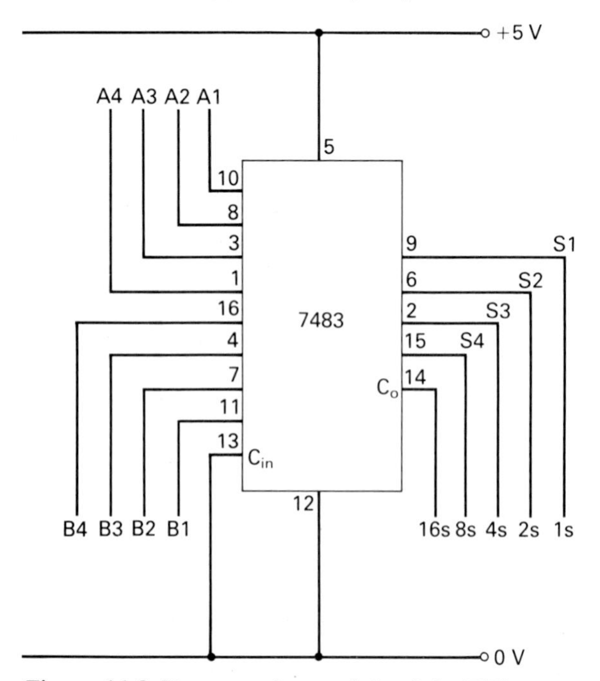

Figure 14.6 Pin connections and circuit for TTL and CMOS adders

When the two 4-bit numbers, 1111_2 and 1111_2 are added (decimal 15 + 15), these two 4-bit adders give the result 11110 (decimal 30). The most significant bit, i.e. the '16s' bit, is available at the carry out (CY or C_{out}) output. The carry in (C_I or C_{in}) input is connected to 0 V. Two 8-bit numbers can be added by using two of these 4-bit full-adders.

14.5 *Experiment* E10

Designing a 4-bit adder

Figure 14.7 shows the connections required to make a practical 4-bit adder. The pin numbers in brackets apply to the 4008, and those without brackets to the 7483. If the CMOS device is used, inputs A_1 to A_4 and B_1 to B_4 should be tied down to 0 V with 100 kΩ resistors to prevent them 'floating' when not taken logic 1 by the switches.

The 4-bit numbers to be added are most conveniently obtained using binary-coded rotary switches, SW_1 and SW_2. These have four binary-weighted output terminals on which the 4-bit word appears, and a common terminal which is connected to + V. The switches are operated by a screwdriver. Alternatively, but less conveniently, wire links could be used to set the input data. Note that the carry-in (C_{in}) terminal is connected to 0 V. The 5-bit result of the addition is evident on the LEDs. Thus, if the two numbers 1111_2 (decimal 15) and 1111_2 are added, the result will be 11110_2 (decimal 30). Of course, by taking the carry-in pin to +V an additional input bit is provided and the result will be 11111_2 (decimal 31). The breadboard layout shows the connections required for the 4008 full-adder.

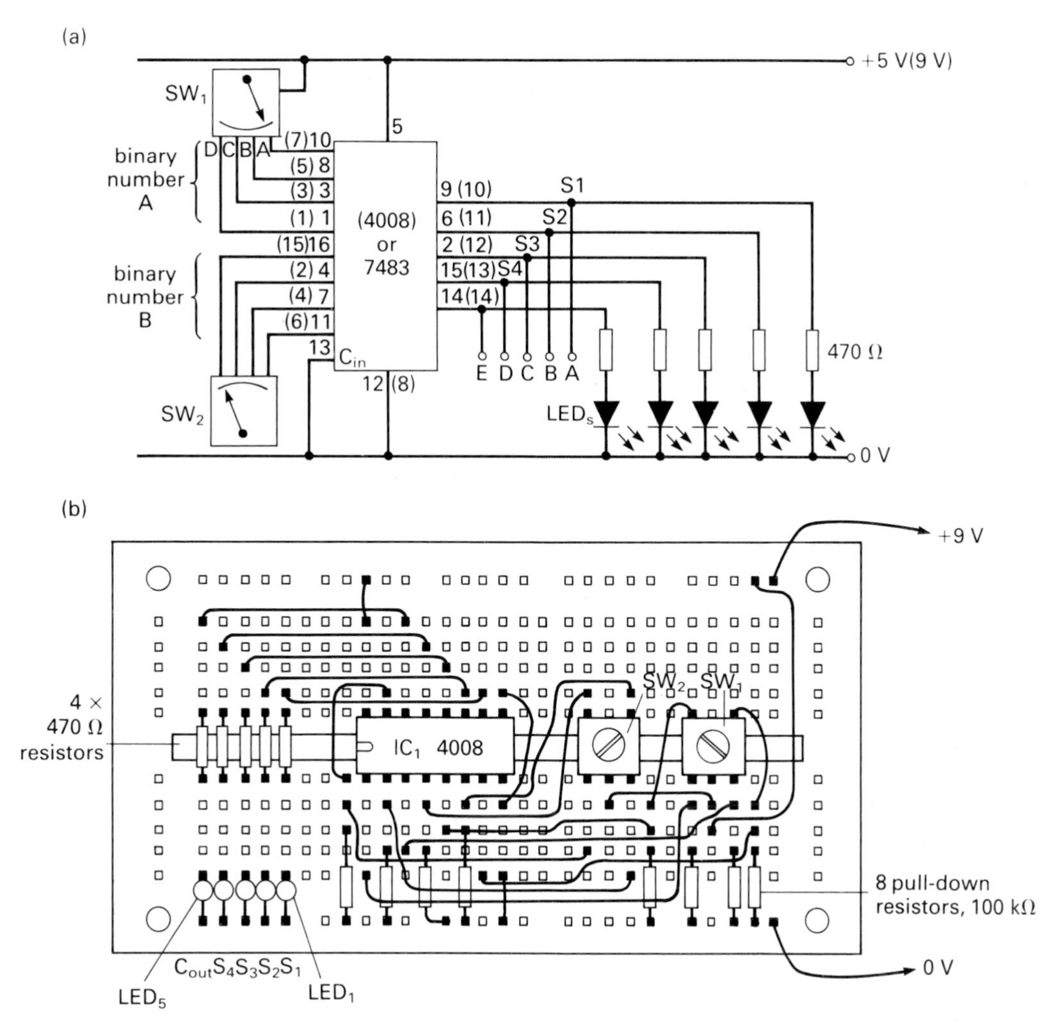

Figure 14.7 A 4-bit full-adder using the 7483 device or 4008 device: (a) circuit diagram; (b) breadboard layout using the 4008

14.6 Magnitude comparators

These are digital comparators (as opposed to analogue comparators using op amps — see Book D, Chapter 4) which are used in circuits to determine if two binary numbers are equal, or which has the greater magnitude.

Figure 14.8 shows the 7485 TTL and the 4585 CMOS 4-bit magnitude comparators. These comparators make fully decoded decisions about two 4-bit words. They can also be expanded to any number of bits without extra gates by connecting two or more of them in cascade using the cascaded inputs.

Thus in the circuit for the 4585 shown in figure 14.8, suppose the binary count on the B inputs is 1011_2 (decimal 11). If the count on the A inputs is also 1011_2, the $A = B$ output (pin 3) goes high, and the $A < B$ (pin 12) and $A > B$ (pin 13) are low. If the count on the A inputs increases to 1100_2 (decimal 12), the $A > B$ output goes high and the $A = B$ and $A < B$ outputs are low. Similarly if the count on the A input is 1010_2 (decimal 10), the $A < B$ output is high and the $A > B$ and $A = B$ outputs are low.

Section 18.5 describes the construction and uses of Project Module E4, a 4-bit Magnitude Comparator, which makes use of the 4585 device.

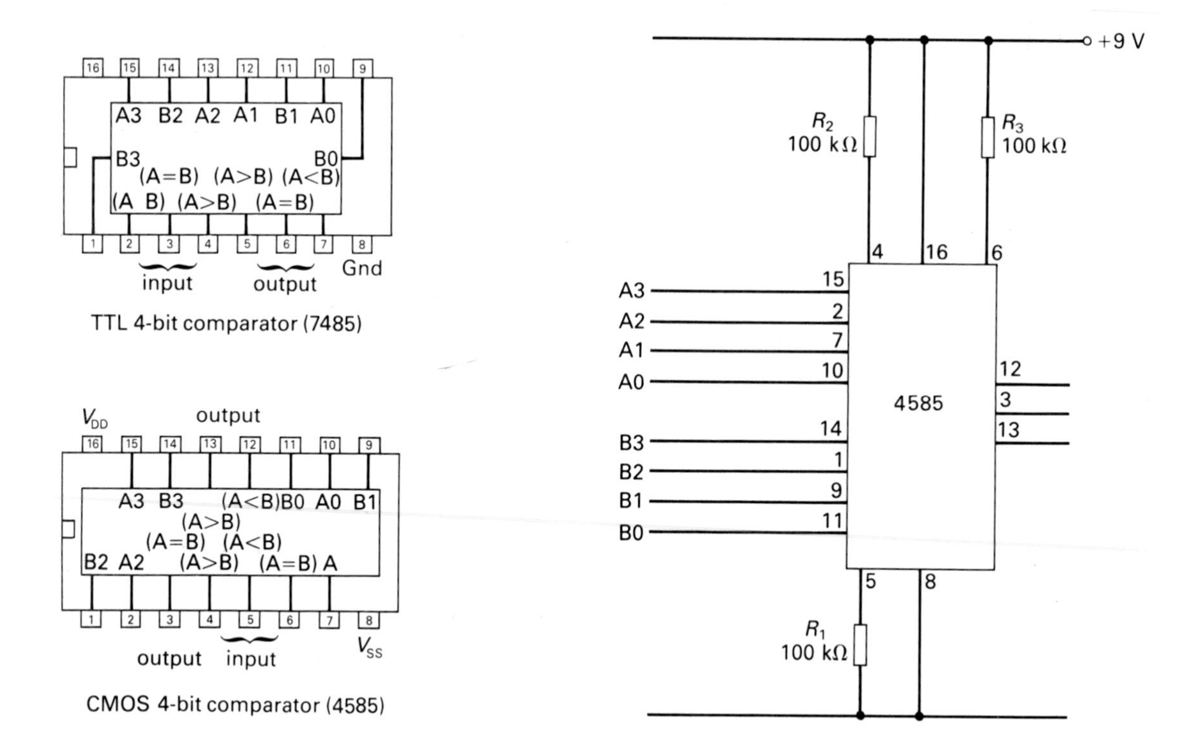

Figure 14.8 Pin connections and circuit for TTL and CMOS 4-bit comparators

15 Memories

15.1 Introduction

An electronic memory is a device that stores information away for future use, in an analogous manner to human memory. This information is stored in an electronic memory as a collection of binary digits (or bits), i.e. 1s and 0s. Memory is therefore a feature of digital systems and not of analogue systems. The information, or data, is usually stored in digital memory as groups of bytes. Microcomputers, calculators, electronic games, and an increasing number of other digital devices make extensive use of memory.

The way memory is used in microcomputers is shown in the simple diagram of figure 15.1. Here data flows between the memory devices and a central processor unit or CPU (also called a microprocessor), and between input and output devices. In a computer system such as this, memory is categorised either as random-access memory (RAM), or as read-only memory (ROM). Both types of memory are made from thousands of transistors formed as an integrated circuit on a small chip of silicon as explained in Chapter 16. In addition to RAM and ROM semiconductor-storage memories, data may be stored external to the computer system as bulk-storage memory which includes magnetic tape and disk.

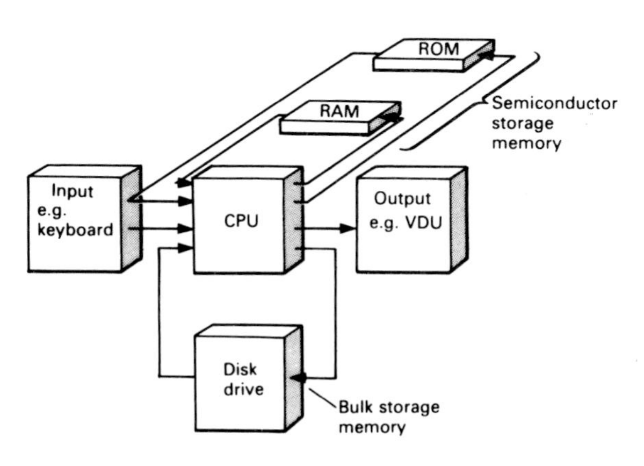

Figure 15.1 The basic microcomputer system

15.2 Random-access memories (RAMs)

The RAM-type of memory can be made to 'learn' — a process called 'writing information into it'. And the data which it remembers can be recalled at any time — a process called 'reading data from it'. Some people prefer to call the RAM a 'readily-alterable memory' since it is a device whose contents can be altered quickly and easily. With very few exceptions, RAMs lose their contents when the power is removed and are thus known as 'volatile' memory devices. All microcomputers use RAM to store data and programs written (or loaded) into it from a keyboard, or from an external store of data such as magnetic tape or disk.

RAMs are often described in terms of the number of bits, i.e. 1s and 0s, of data that they can hold, or in terms of the number of data words, i.e. groups of bits, they can hold. Thus a 16 384 bit RAM can hold 16 384 1s and 0s. This data could be arranged as 16 384 1-bit words, 4096 4-bit words, or 2048 8-bit words. Semiconductor memories vary in size, e.g. 4K, 64K, and 128K. Here we are using K defined as:

$$K = 2^{10} = 1024.$$

Thus a 16K memory has a storage capacity of $16 \times 1024 = 16\ 384$ words, a 128K memory holds 131 072 words, and so on.

There are two main members of the RAM family: static RAM and dynamic RAM. The essential difference between them is the way in which bits are stored in the RAM chips. In a static RAM, the bits of data are written in the RAM just once and then left until the data is read or changed. In a dynamic RAM, the bits of data are repeatedly rewritten in the RAM to ensure that the data is not forgotten.

15.3 Static RAMs

Flip-flops (Chapter 5) are the basic memory cells in static RAMs. Each flip-flop is based on either two bipolar transistors, as shown in figure 15.2, or on two metal-oxide semiconductor field-effect transistors (MOSFETs). As many of these memory cells are needed as there are bits to be remembered. Thus in a 16K-bit static memory, there are 16 384 flip-flops, i.e. 32 768 transistors. All these transistors would be accommodated on a single silicon chip about 4 mm square.

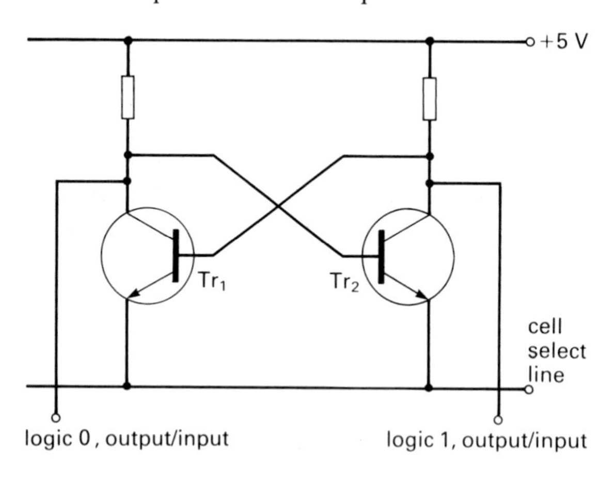

Figure 15.2 The flip-flop memory cell in a static RAM

The 7489 TTL device shown in figure 15.3 is an example of a simple static RAM. It has 64 memory cells, each cell capable of holding a single bit of data. The cells are organised into locations, and each location is capable of holding a 4-bit word. Thus the 7489 is capable of storing sixteen 4-bit words, i.e. four cells are used at each location. Each location is identified by a unique 4-bit address so that data can be written into or read from these locations. Note that the number of words stored in memory determines the size of the address word. Thus the number of address SELECT lines (pins 1, 15, 14, and 13) is four, since $2^4 = 16$. There are four data input lines (pins 4, 6, 10, 12) on which data is placed to be stored in memory and there are four data output lines on which data is read out from the RAM.

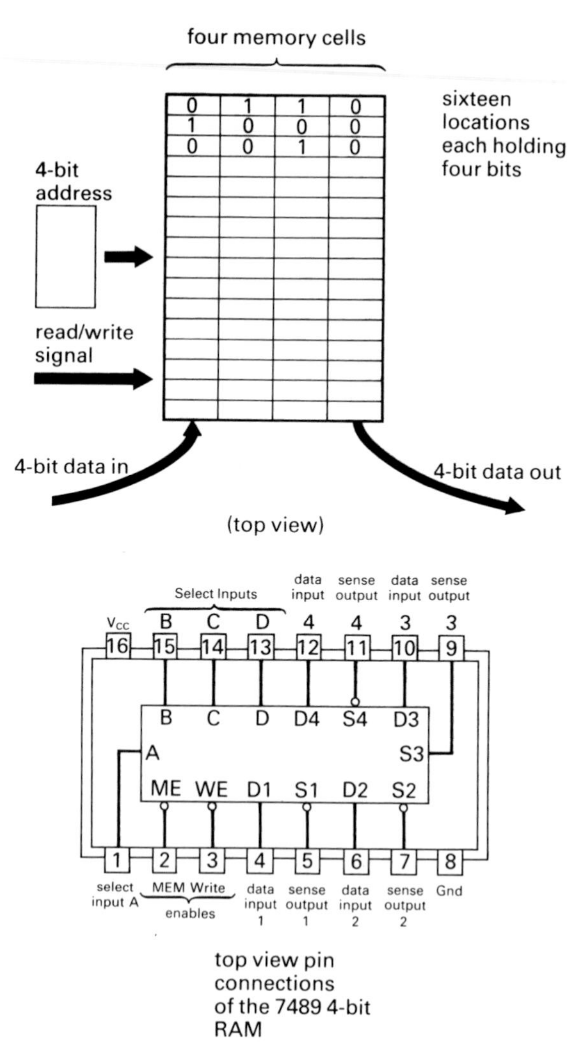

Figure 15.3 Memory organisation of the 7489 static RAM

15.4 Dynamic RAMs

Unlike a static RAM which holds its data until 'told' to change it, a dynamic RAM continually needs to have its data refreshed. Dynamic RAMs (also called DRAMs) are based on metal-oxide semiconductor field-effect transistors (MOSFETs — see Book C, Chapter 25). Bits of data are stored in dynamic RAMs as small packets of charge, rather than as voltage levels as in the static RAM. This has the advantage that the power consumption of MOSFET memory circuits is very low.

As shown in figure 15.4, each memory cell in a DRAM is a very simple circuit and comprises a small capacitor and a single n-channel MOSFET which is switched on to read this charge. Each cell holds a 1 bit as a tiny electrical charge of about 10^{-15} coulombs. Though tiny, this charge still amounts to about 5000 electrons! However, the charge on the capacitor tends to leak away and extensive 'refresh' circuitry is needed to keep the charge 'topped up'. The additional electronics required to ensure that a dynamic RAM doesn't forget what's in it adds to the cost and complexity of dynamic RAMs. However, the newer dynamic RAMs have refresh circuitry on the chip with the memory cells.

Dynamic RAMs tend to be cheaper than static RAMs for large-capacity memory devices. This is because of the smaller size of the dynamic RAM cell as opposed to the static RAM cell, as it is based on one transistor instead of two. The first 1K-bit DRAM was introduced in the early 1970s, and since then the number of cells on a memory chip has doubled every year, culminating in the latest 1M-bit devices, and the proposed 4M-bit memories by 1989. Today the market for memories accounts for over half the total market of integrated circuits.

15.5 Read-only memories (ROMs)

The problem with a random-access memory is that its memory is volatile, i.e. it loses all its data when the power supply is switched off. A non-volatile memory is a permanent memory that never forgets its data. One type of non-volatile memory is the read-only memory (ROM). A ROM has a pattern of 0s and 1s imprinted in its memory by the manufacturer. It is not possible to write new information into a ROM, i.e. it is a 'read-only' memory.

The organisation of data in a ROM is similar to that of a RAM. Thus a 256-bit ROM might be organised as 32 words each

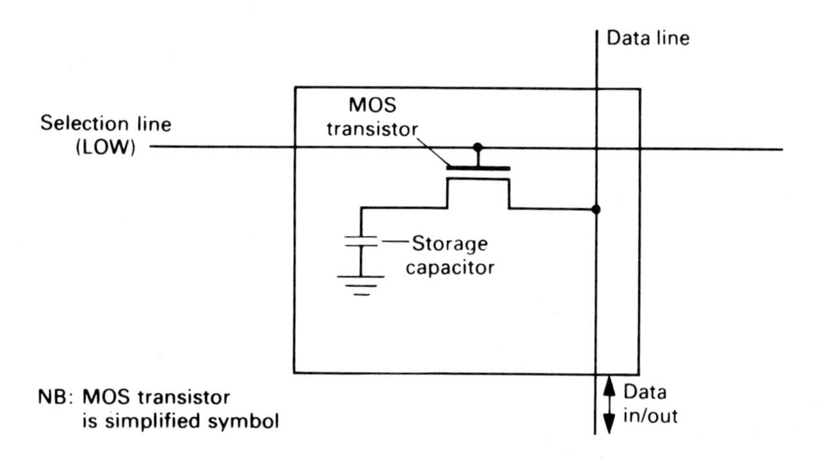

Figure 15.4 A storage cell in a dynamic RAM

8-bits long; a 1024-bit (1K-bit) ROM might be organised as a 256 × 4-bit memory, and so on. Figure 15.5 shows the 7488 256-bit ROM organised as 32 × 8-bit words. Any one of these 8-bit words may be addressed through the five SELECT lines which identify the location of each 8-bit word. When pin 15, the MEMORY ENABLE pin is taken low, the word stored in the location appears on the 8 output lines.

A ROM may be regarded as the 'reference library' of the computer world. For example, microcomputers have an integrated circuit ROM which stores instructions such as the language and graphics symbols the computer uses. A typical ROM for a microcomputer has 8 kilobytes of memory (1 byte is an 8-bit word). Thus it stores 65 536 bits of data on a single chip which is generally contained in a 28-pin DIL package. Through these 28 pins, the microcomputer is able to select and read any one of the 8192 locations in which the bytes of data are stored.

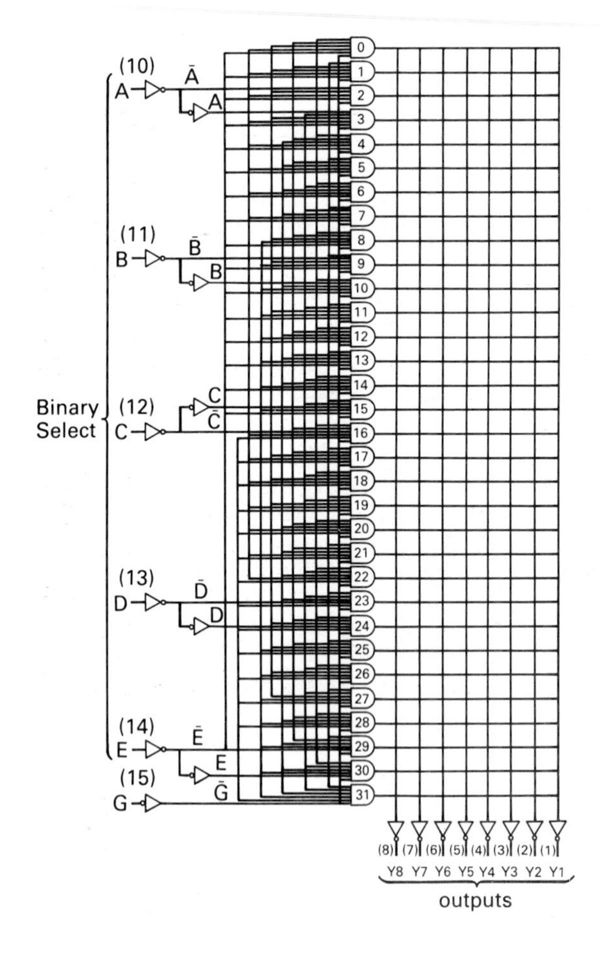

Figure 15.5 An extreme simplification of the 256 program options of a 7488 256-bit ROM (Pin numbers shown in brackets)

15.6 Programmable read-only memories (PROMs)

The memory cells in a bipolar ROM are simply npn transistors which have a fuse link placed in series with the emitter of each transistor as shown in figure 15.6. Suppose a decoder addresses transistor, Tr_1, which is one of eight in a stored 8-bit word, say. With the fuse link in place, Tr_1, will be 'on' and the SENSE line will be pulled high by the addressed cell. This high forward-biases the buffer transistor, Tr_2, thus turning it on. The collector voltage of Tr_2 is then low, i.e. at logic 0. However, if the fuse link is open, the transistor is 'off' and the SENSE line will be pulled low. Hence the collector voltage of Tr_2 is high, i.e. at logic 1. The fuse links are made of some material such as titanium-tungsten or nichrome. These transistor ROM cells are programmed by one of two processes, *mask programming* and *field programming*.

Mask programming is accomplished when a ROM is made. If a '1' is to be stored at a particular memory location, the fuse link of the transistor at that location is simply not connected. The second method of programming a ROM is left to the user. In a field-programmable ROM,

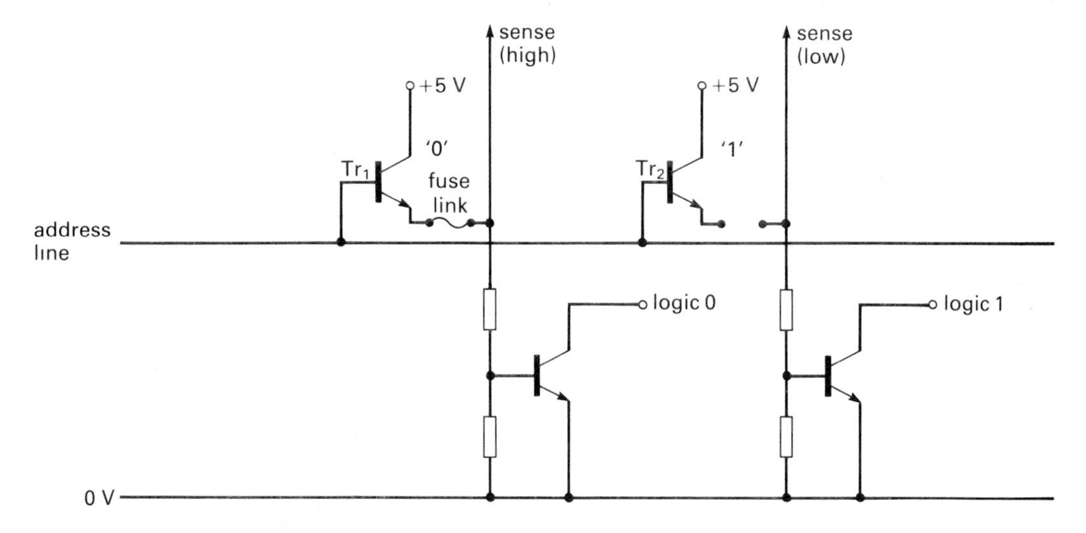

Figure 15.6 The use of a fuse link in a PROM

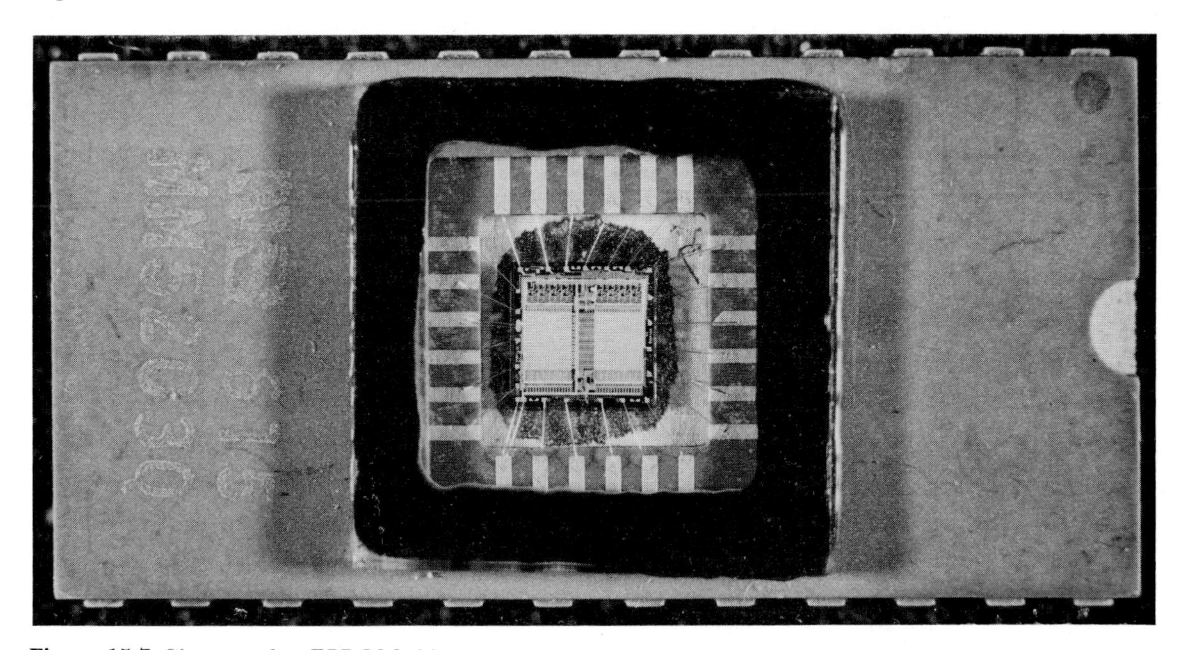

Figure 15.7 Close up of an EPROM chip

i.e. a PROM, the user stores a 1 in a cell by addressing the cell and applying a high current pulse (approximately 30 mA) to it which 'blows the fuse'. This once-only programming is done with a device called a PROM burner. To read the contents of a particular cell, the cell is addressed by an address decoder that is usually an integral part of the PROM chip. ROMs are used for dedicated applications such as controlling the actions of an industrial robot, or providing a microcomputer with a wordprocessing capability.

Both mask and programmable ROMs are rendered useless if there is just one incorrect bit of data stored in the chip. Fortunately, another type of ROM, the EPROM (or erasable-programmable ROM), has a memory which can be erased and reprogrammed many times. A typical EPROM is shown in figure 15.7 and is based on MOSFET transistors. The quartz window over the top of the chip allows a beam of ultraviolet light to erase any stored data in the chip. This usually takes about 10 minutes. A fresh set of instructions is written in the EPROM using an EPROM programmer which re-establishes a 'packet' of charge in selected MOSFET transistors; the charge remains there indefinitely unless erased by UV light.

15.7 Magnetic bubble memories (MBMs)

Bubble memories store bits of data on small magnetised domains called 'bubbles'. The domains move in a thin film of magnetic garnet which is deposited on a non-magnetic garnet substrate. It is possible to destroy bubbles to clear the store, generate bubbles in order to write new data, and replicate bubbles for reading out the data. Stored data can be accessed by moving the bubbles along a closed track called a 'major loop' which links the write, read and erase stations as shown in figure 15.8.

The track along which the bubbles are moved is defined by patterns of permalloy deposited on a thin magnetic film grown on to a substrate of gadolinium-gallium-garnet. The bubbles are created by electrical signals that circulate in very small conducting loops just above the film. The bubbles move under the control of a rotating magnetic field. Bubbles are detected when they pass under the permalloy strips deposited on the bubble-bearing film. The magnetisation of the strips, and hence their resistance, changes due to the presence of a bubble (which represents a 1). The absence of a bubble represents a 0.

Data is transferred from the major loop to minor loops for storage; it is accessed by transferring data from the minor loops after these have been rotated so that the required data is next to the major loop. By using a number of minor loops, the time to access data can be reduced considerably since one bit from each of the minor loops can be transferred in parallel to the major loop.

The main advantage of an MBM is that it is non-volatile (like a ROM) since the

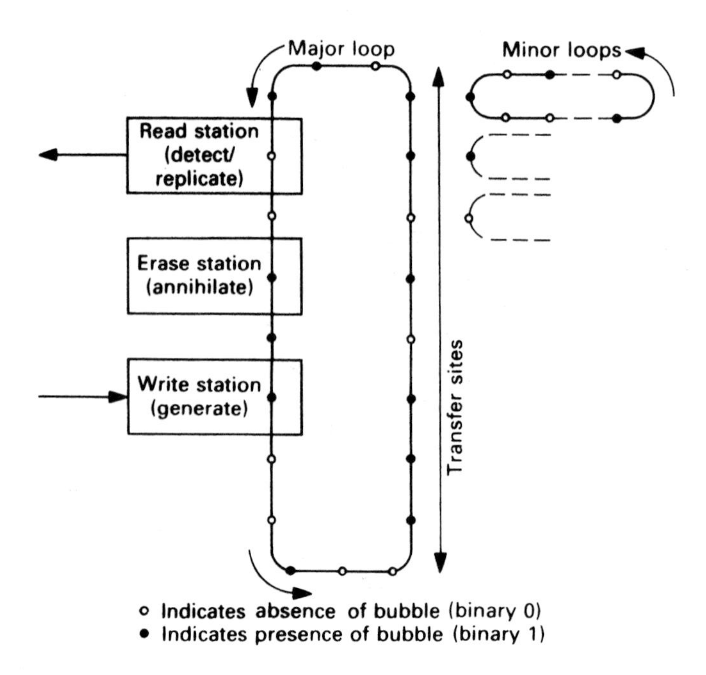

Figure 15.8 A simplified schematic of a magnetic bubble memory

bubbles do not disappear when the power supply is switched off. However, unlike a ROM, an MBM can have data written into it at any time. MBMs are quiet and very reliable since they have no moving parts. And since an MBM can store over a million bits in a very small volume, it is suitable for a whole range of microprocessor-based products such as wordprocessors, data loggers and microcomputers.

15.8 Floppy disks

Another type of bulk storage device on which large amounts of data can be stored is the floppy disk (also called diskette, or mini-disk). The data is written to and read from floppy disks using a disk drive — see figure 15.1. There are several sizes of floppy disks but the one commonly used with microcomputers is 5.25 inches across. It is made of flexible Mylar film which is coated with a thin magnetic film on which binary data is stored in the form of tiny magnetised regions. A 3.5 inch disk is also available.

As shown in figure 15.9(a), a floppy disk is enclosed within a plastic jacket. There is a round hole in the middle of the jacket which enables the hub of the disk drive to clamp on the disk and rotate it at 300 rev/min. Data is stored on and retrieved from the disk by the read/write head of the disk drive which touches the spinning disk through the oblong hole at the edge of the disk. Note that since the read/write head can move across the surface of the disk, the floppy disk (in common with the compact disc used for storing audio information) is a *random/access* device. This

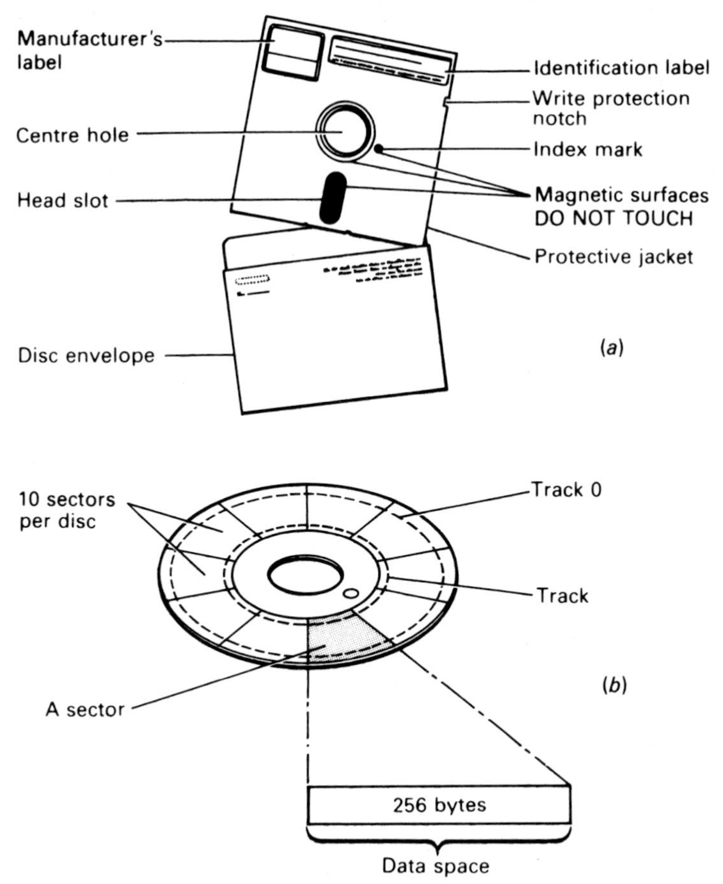

Figure 15.9 A floppy disk: (a) its general features; (b) its tracks and sectors

means that the read/write head of the disk drive (and the laser head on a compact disc) can jump to any location on the disk. Thus the data on a floppy disk can be accessed much faster than data on a magnetic tape. The tape player has to sequence through a lot of data on the tape before locating a certain section, i.e. it is a *serial access* device.

The data on a floppy disk is organised into concentric circles called tracks (figure 5.9(b)). There may be forty tracks on a disk and each track is divided into ten sectors. Each sector is further divided into 256 bytes of data where one byte (an 8-bit word) represents one stored character. A 'single-sided, single-density' disk has 40 tracks on one side only; a double-sided, double-density disk has 80 tracks on each side. Thus a single-sided single density 5.25 inch floppy disk can store 40 tracks × 10 sectors × 256 bytes = 102 400 bytes.

The disk filing system of a computer automatically records the location of data, e.g. programs, on a disk. The first two sectors on track zero of a disk are reserved for this purpose — they hold the 'catalogue' of the disk. Whenever a piece of data is required from the disk, the disk filing system first reads the catalogue to find out which tracks and sectors are used to store this data. Clearly, before a floppy disk can be used, it has to have the tracks and sectors created on the disk so that the disk filing system can find the data required. This process is called 'formatting' a disk and includes setting up the tracks and sectors and creating the catalogue. A formatting routine is usually provided with a microcomputer to enable floppy disks to be formatted, but it should be remembered that there is no standard way of formatting a floppy disk.

A typical sector format is shown in figure 5.9(c). Each sector is divided into data fields. When the disk drive rotates the floppy disk, the address mark passes the read/write head first and identifies the next area of the sector as the ID field. The ID field identifies the data field by sector and track number, and the data mark indicates whether the upcoming data field contains a current or active record, or a deleted record. The data field is that part of the sector that contains the stored data.

15·9 CCD (charge-coupled device)

Light dependent resistors, solar cells and phototransistors respond to changes of light intensity. And so does a type of sensor shown in figure 15.10 which is called a charge-coupled device (CCD). This is a novel data storage device which consists of an array of semiconductor memory cells on a silicon chip that respond to light by storing a small charge. This charge can be removed from the array and processed to form an image by electrodes attached to the surface of the chip. The longer light falls on the CCD, the more charge is stored. Thus the CCD is very sensitive and can be used to gather light from very faint sources. In telescopes, a CCD is more sensitive than a photographic plate and enables light from faint stars and galaxies to be detected (figure 15.11), and in TV cameras CCDs can be used for security and militiary purposes since images can be obtained in near darkness. CCDs are also used in the field of robotics since a CCD can recognise a range of grey tones. They have also found their way into the ubiquitous 'cam-corders.'

The way a charge-coupled device works is shown in figure 15.12. It provides a cross-section view of the semiconductor chip making up the CCD. On the surface of the chip is a long row of tiny metal spot (electrodes) which overlay a thin oxide layer formed on the surface of a p-type substrate. A three-phase clock network alternately activates the electrodes in turn by being switched from 0 V to say +10 V. When an electrode is pulsed to a positive voltage, it is capable of attracting a negative charge to the underside of the

Figure 15.10 Charge-coupled device (CCD) Courtesy: English Electric Valve Co Ltd

Figure 15.11 Star clouds in the constellation of Sagittarius, which lies towards the centre of our galaxy, the Milky Way. The bright 'star' at the bottom is the planet Jupiter. The glow to the right is the Lagoon Nebula. The CCD camera can give more detail than optical photos like this.

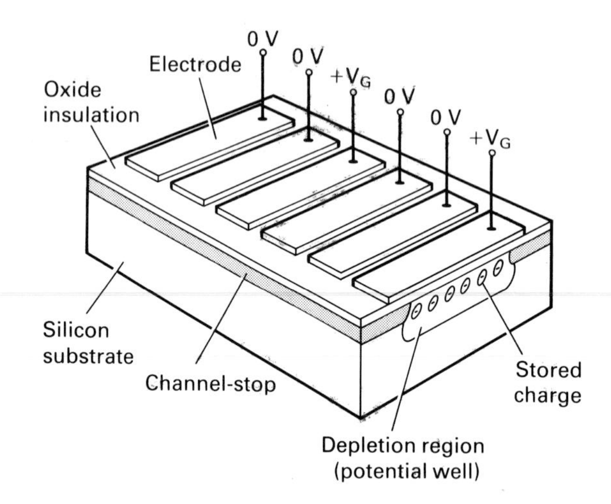

Figure 15.12 Cross-section through a CCD

oxide layer beneath it. It is as if the positively-charged metal electrode creates a kind of electrical 'bucket' that can hold electric charge.

'Charge-coupling' is the technique by which signal charge can be transferred from the bucket under one electrode to the next bucket. This is achieved by taking the voltage on the second electrode also to 10 V then reducing the voltage to 0 V on the first electrode as illustrated in figure 15.13(a). Hence by sequentially pulsing the voltages on the electrodes between high and low levels, charge signals can be made to pass down an array of very many electrodes. To achieve this the electrodes are connected in sequence to a set of three-phase drive pulses as illustrated in figure 15.13(b). Charge signals can then be stored under every third electrode in the array and will be transferred together along the array under control of the drive pulses. The use of three phases rather than two ensures that charges move in the right direction. In this respect, the operation of the CCD is like a shift register (Chapter 13). By letting the presence or absence of a charge represent digital values of 0 and 1, and by providing amplifiers for injecting and detecting these charges, a very simple and compact type of computer memory device is possible.

When a CCD is used as an electronic imager in a camera, the metal electrodes are overlayed by optically-sensitive surfaces. For example, when used for 625-line TV operation, the image area is a 385 × 288 array of pixels covering the same number of charge storage sites used in a TV camera for detecting light.

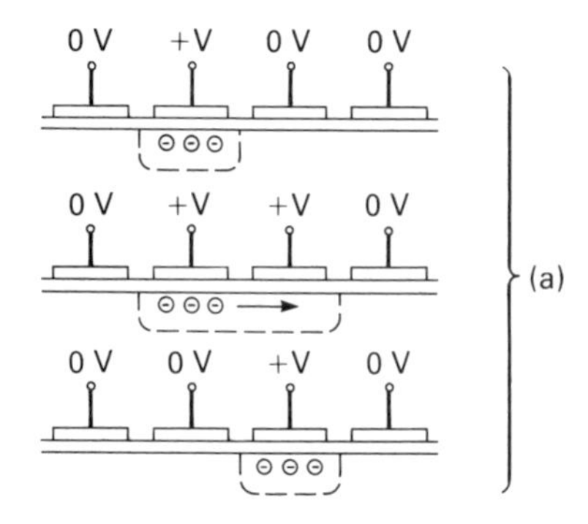

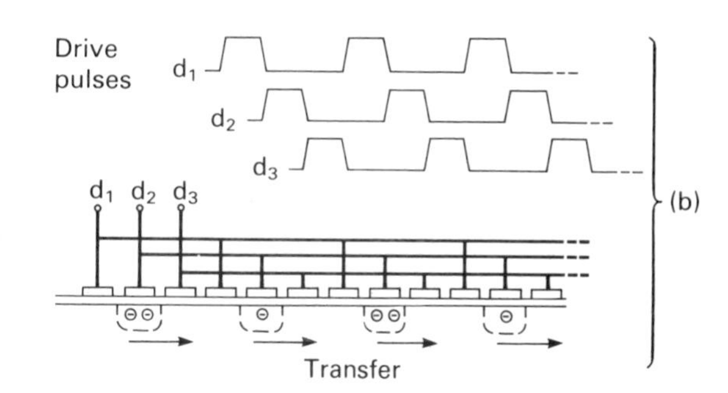

Figure 15.13 How the charge is transferred in a CCD

15.10 CD-ROM

The CD-ROM (*c*ompact *d*isc-*r*ead *o*nly *m*emory) is one medium which has enormous potential in the immediate future as a computer peripheral for the mass storage of data. Based on the format of the highly successful audio compact disc (Book A, Chapter 1) which can store 60 minutes of digital audio, each 12 cm diameter CD-ROM can hold 543 megabytes (MB) of user data though it is possible to store 650MB if the outer regions of the disc are used. And that's equivalent to 150000 pages of ASCII text at 2000 characters per page, or 15 000 pages of images at 34KB per page, or 3000 colour pictures at 180KB per picture, or 1500 floppy discs!

Moreover, the manufacturers of audio compact discs and CD-ROMs use the same technology so that the same manufacturing plant can share economies of scale. However, the correcton for errors on such discs does lead to one major difference in the two systems. In an audio system, an incorrectable error such as the loss of bits due to surface dust, can simply be muted and is rarely noticeable. But if data is being read from a CD-ROM at, say, 1Mbits per second, an error rate of, say, 10^{-6} provides a quite unacceptable one error per second in the data flow. Nevertheless, the use of CD-ROMs in the computer field seems assured though they do not have the one advantage that a floppy disc has – they can only be read to not written to as well.

However, there are 'write once' versions of the CD-ROM (colourfully known as WORMs – *w*rite-*o*nce, *r*ead *m*any times). So users can store their own data once only. The surface of these discs consists of an extremely thin metal film deposited in a vacuum. A powerful laser is used to heat a pinpoint of metal rapidly to beyond its melting point so that it forms a tiny crater on the surface in cooling. These surface features are used to represent data. And of course, like an ordinary audio CD, a considerably less powerful laser is used to read the data.

CD-ROMs are already in use in libraries as stand-alone look-up devices for encyclopedias and other reference works, as a device for interactive training and education, and as a multimedia device for text mixed with graphics and sound. The latter application is known as CD-I meaning *c*ompact *d*isc-*i*nteractive for which there is a growing interest as a low-cost alternative to interactive video (IV) which relies on a costly videodiscs such as Laservision.

15.11 Access times

The access time for a particular memory is the time it takes to locate and deliver a piece of data in the memory. RAMs and ROMs (semiconductor storage memories) have faster access times than disks, tapes and bubble memories (bulk-storage memories). Thus RAMs have an access time of less than 1 μs; e.g. the 6116 16 K-bit MOSFET static RAM has an access time between 100 ns and 250 ns. Bipolar RAMs have faster access times than this. ROMs have slower access times than RAMs. Of the bulk storage devices, the floppy disk (a parallel access memory) has an access time of about 30 ms. Slower still is magnetic tap (a serial access device) with an access time of several seconds or even minutes. MBMs have access times of between 10 and 100 μs but a short access time is not all important and has to be traded against the cost per bit stored in a memory.

15.12 Gallium arsenide

Gallium arsenide (GaAs) is a crystalline substance which, like silicon, is used to make diodes, transistors and integrated circuits.

Gallium arsenide's main claim to fame is that semiconductor devices made from it conduct electricity five times faster than those made from silicon. And this property makes gallium arsenide an interesting material to weapons manufacturers since data in computer memories made from GaAs can be accessed more quickly than memories made from silicon. This means that missiles and other weapons can respond rapidly to sensing and control circuits.

However, there are a few drawbacks to the use of GaAs, one of which is that the two elements gallium and arsenic from which it is made are in short suppy, mainly being found as impurities in aluminium and copper ores. On the other hand, silicon is plentiful being found in silicates such as sand. The cost of GaAs is about thirty times that of silicon, and this is aggravated by the fact that about 90% of GaAs chips are rejected after production. Furthermore, it is not as easy to make integrated circuits from GaAs as it is from silicon since it does not form a protective layer of oxide to resist the diffusion of dopants during the process of photolithography — see Chapter 16.

It is therefore unlikely that there will be an increase in the use of GaAs-based semiconductors in the near future except for specialist applications where cost is not a major consideration, e.g. ballistic missile development and the USA's 'Star Wars' programme, the Strategic Defence Initiative (SDI). But manufacturers of GaAs have turned to space for help in overcoming the problems and costs of producing pure GaAs. A number of countries are planning to produce pure crystalline GaAs in the extremely low vacuum and zero gravity in laboratories aboard orbiting space stations.

16 Making Integrated Circuits

16.1 Introduction

An integrated circuit is made of microscopically small transistors, diodes, resistors and other components connected together on a 2 mm to 5 mm square chip of silicon. The integrated circuit produced in this way is a complete audio amplifier, or analogue-to-digital converter, or memory, or any other function for which the IC was designed. The components making up these ICs are the smallest man-made objects ever created and the trend is towards producing even smaller components on a silicon chip.

16.2 Moore's Law

In the 1960s, George Moore, founder of the Intel Corporation in the USA, said that the number of components that would be integrated on a single silicon chip would double every year (figure 16.1). To the present day, Moore's Law (as it came to be known) has been found to hold good. In Moore's day, ICs were made from a few tens of components. By the late 1960s, the annual doubling effect had led to several hundred components being integrated on a silicon chip (known as medium scale integration — MSI). During the 1970s, component counts per chip had reached several hundred thousand (known as large scale integration — LSI). In the 1980s and 90s, microprocessor and memory chips now have close to a million components on a single silicon chip, and this is called very large scale integration — VLSI. The small size of components on a silicon chip is illustrated by a simple comparison: if a present-day 5 mm square memory chip were the size of the UK, the smallest components would be the size of a tennis court.

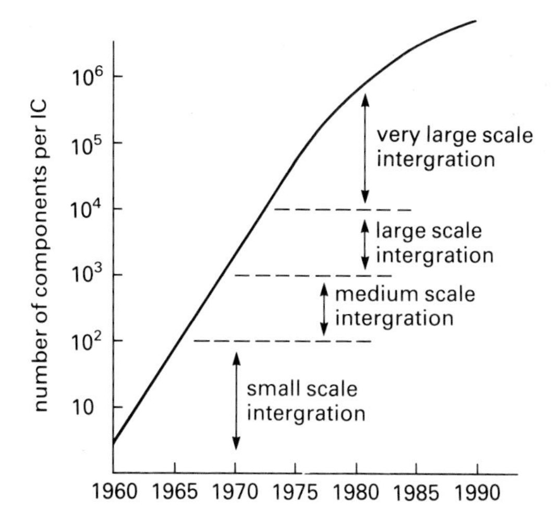

Figure 16.1 Graph showing increase in chip complexity

16.3 Photolithography

The process of putting several hundred thousand transistors on a silicon chip a few millimetres square is not easy. A technique called photolithography is at the heart of the process. Photolithography uses photographic techniques and chemicals to etch a minutely-detailed pattern on the surface of a silicon chip. Each stage in the process involves the use of photographically-prepared plates called photomasks. Each photomask (figure 16.2) holds a particular pattern identifying individual transistors, conducting pathways, etc. Photomasks are produced by the photographic reduction of a much larger pattern. A photomask is placed over a thin layer of photoresist covering the surface of the silicon. Ultraviolet light

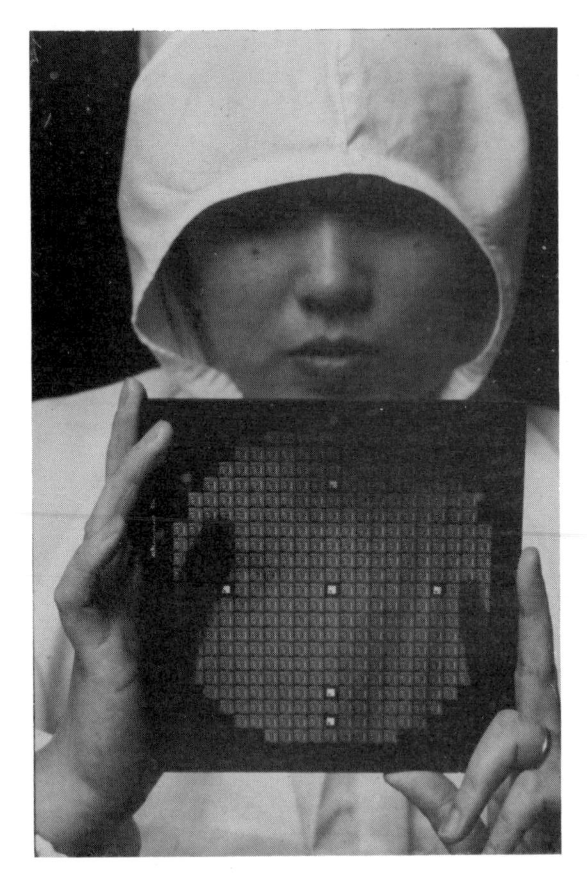

Figure 16.2 A scientist examines a photomask

directed on the photomask passes through the clear areas but is stopped by the opaque areas. According to the type of photoresist used, either the exposed or the unexposed photoresist can be dissolved away with chemicals to leave a pattern of lines and holes. This pattern enables transistors to be formed in the silicon, and allows aluminium interconnections to be made between them.

16.4 Preparing a silicon wafer

The process of making a silicon chip begins with a 50–150 mm cylinder-shaped single crystal of pure silicon (or gallium arsenide) known as a boule or ingot (figure 16.3). The ingot is obtained by slowly pulling the growing crystal from a bath of pure molten silicon. The ingot is then cut up into thin slices, known as wafers, about the size of a beer mat and half as thick.

Figure 16.3 A boule or ingot of silicon

The wafers are then passed through an oven containing gases heated to about 1200°C. The gases diffuse into each wafer to give it the properties of a p-type or an n-type semiconductor depending on the gas used — a process known as epitaxial growth. The wafers are then ready to have integrated circuits formed on them by a complex process which involves masking, etching, and diffusion.

16.5 Creating windows in the silicon wafer

Figure 16.4 shows the several stages required to produce openings in the surface of the silicon through which gases are diffused to create transistors. The first stage involves heating the wafer to about 1000°C in a stream of oxygen so that a thin layer of silicon dioxide is formed over the whole surface of the wafer.

In the next stage, a thin layer of a light-sensitive emulsion (photoresist) is spread over the layer of silicon dioxide. A photographic plate (the photomask) is placed over the top of the emulsion. The photomask contains a pattern of microscopic dots, which are to become

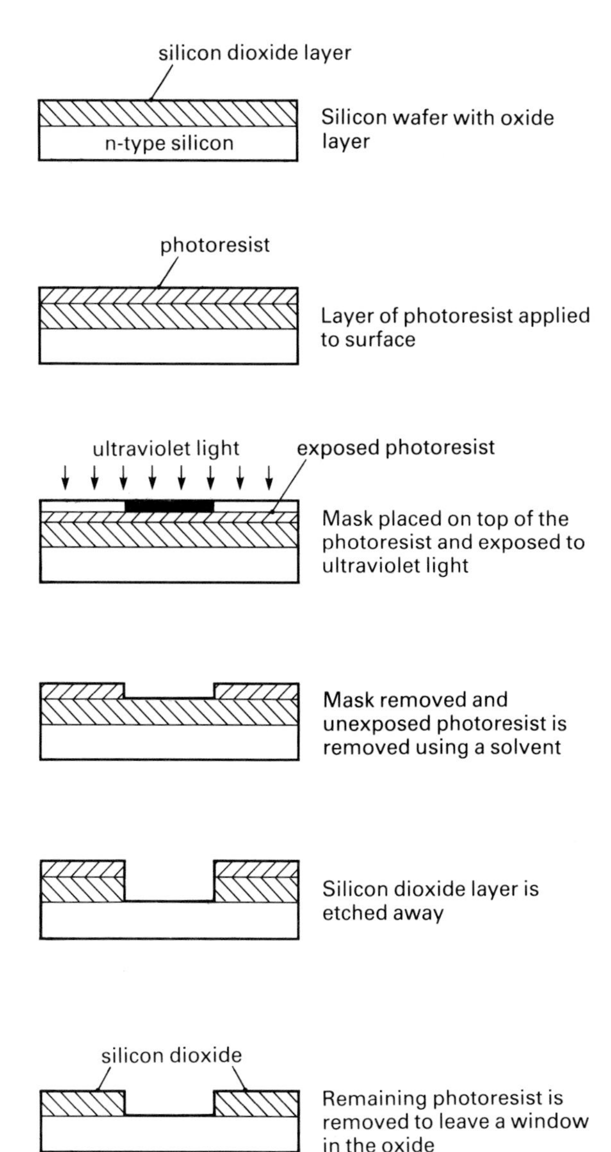

Figure 16.4 Steps in the formation of a window in the silicon dioxide surface of a silicon chip

holes in the silicon dioxide layer. A single mask holds the pattern for several hundred integrated circuits for each wafer.

In the third stage, the mask is exposed to ultraviolet light. Where the mask is transparent, the light passes through and chemically changes the photoresist underneath so that it hardens. The unexposed photoresist can easily be removed with a suitable solvent. This is the fourth stage.

In the fifth stage, the silicon wafer is immersed in another solvent which removes the silicon dioxide from the unexposed areas. The wafer now has a thin surface layer of silicon dioxide in which there are a large number of minute 'windows' — see figure 16.4. It is through these windows that gases are allowed to pass into the epitaxial silicon layer underneath to form transistors. In the production of complete silicon chips on a wafer, the formation of a silicon dioxide layer, followed by masking and etching, has to be repeated many times.

16.6 Making individual transistors

Figure 16.5 shows the various steps required to make an npn transistor on a silicon chip, and figure 16.6 shows that similar steps are required to create a single n-channel MOSFET.

First a gas is selected which diffuses through a window to form a p-type base region in the n-type silicon epitaxial layer. Next a fresh silicon dioxide layer is formed over the window, followed by a stage of masking and etching to create a second smaller window. Through this window a gas diffuses to form the n-type emitter region. Another layer of silicon dioxide is formed over this window followed by masking and etching to create smaller windows for making contacts to the base and emitter regions. These contacts are made by depositing aluminium in vapour form. In the final stages of making an integrated circuit, vaporised aluminium is allowed to form a thin layer of aluminium over the entire surface of the silicon chip. This thin layer is cut into a pattern of conducting paths using the techniques of masking and etching. When all the integrated circuits have been formed in this way, the wafer is cut up into individual chips, checked and packaged in a form which can be used by the circuit designer. The IC package is usually the familiar dual-in-line type.

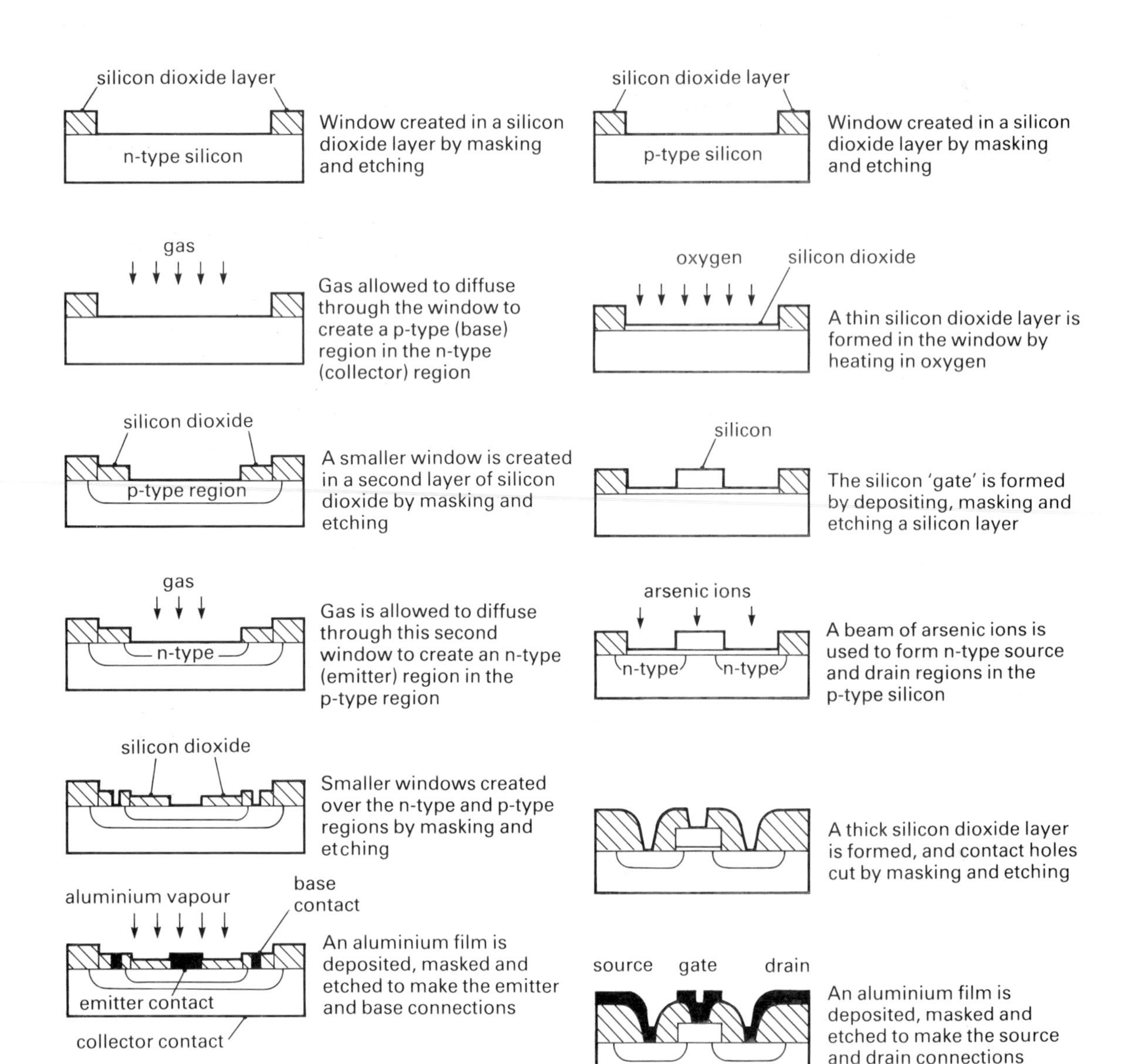

Figure 16.5 Steps in the formation of a single npn transistor on a silicon chip

Figure 16.6 Steps in the formation of a single n-channel MOSFET transistor on a silicon chip

17 Uses for Digital ICs

17.1 Introduction

This chapter describes further practical circuits making use of digial integrated circuits. In addition, remember that this book contains a further seven Project Modules which are described in Chapter 18. And in Book D, Project Modules D5 and D7 are also based on digital ICs.

17.2 Combinational Lock

The circuit is shown in figure 17.1 and enables a solenoid to be switched on by means of an 'electronic key'. If the correct key is inserted, the circuit will latch and the solenoid operates; if an incorrect key is inserted, a warning tone sounds until the correct key is inserted. The insertion of the key also turns on the power to the circuit.

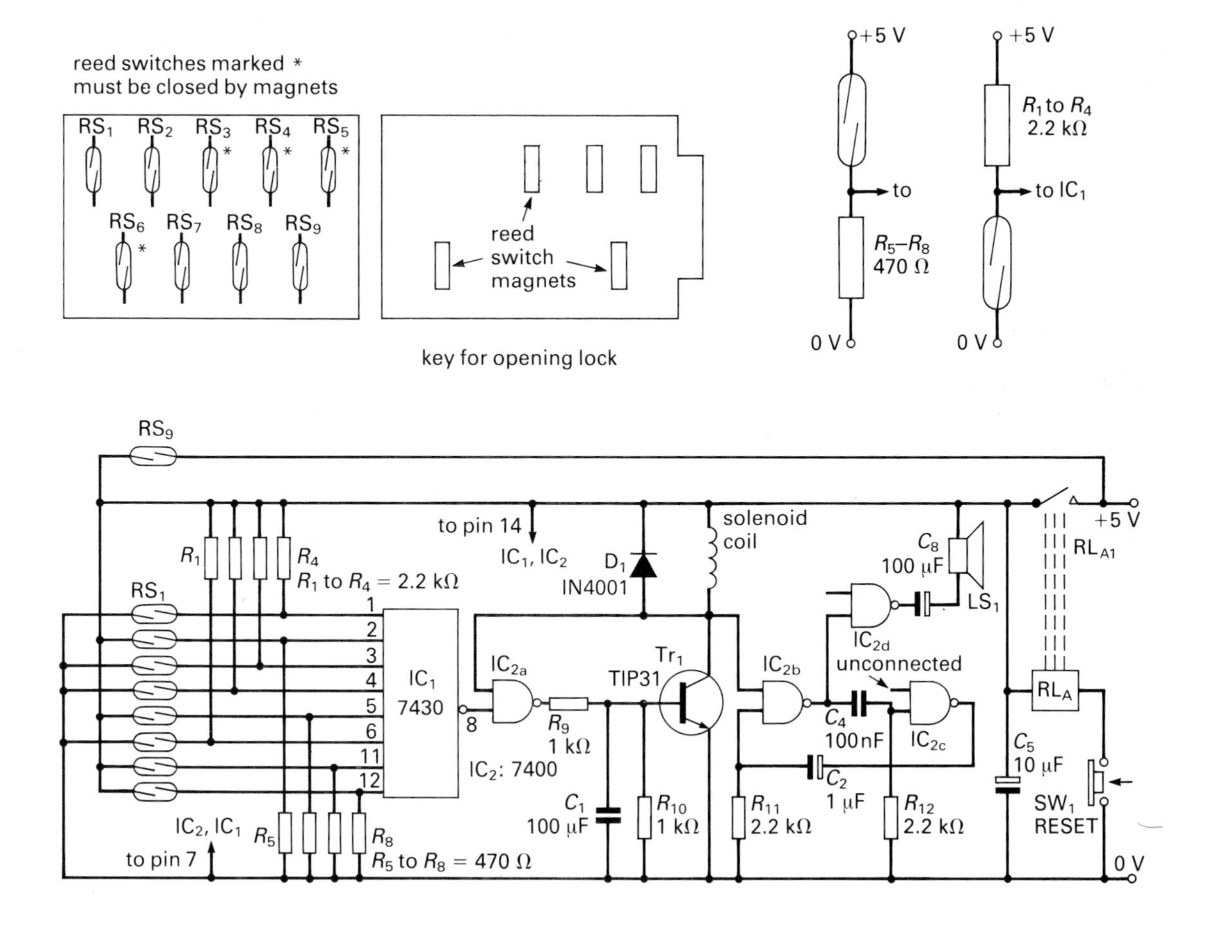

Figure 17.1 Combination lock

An 8-input TTL NAND gate, IC_1, is the main building block in the circuit. If all inputs to the NAND gate are made high by closing the right combination of reed switches, RS_1 to RS_8, the output from the IC_1 goes low. This low output is fed to the NAND gate, IC_{2a}, which forces its output high and turns on Tr_1 which energises the coil of the solenoid. The low signal produced at the collector of Tr_1 is fed back to the input of IC_{2a} which latches the transistor switch. Thus, once the solenoid has been energised and the bolt withdrawn from the lock, the key may be removed. Capacitor C_1 ensures that when the power supply to the circuit is switched on, Tr_1 is off.

The low signal at the collector of Tr_1 disables the astable based on NAND gates IC_{2b} and IC_{2c}. If an incorrect key is used, the collector of Tr_1 is high and the alarm sounds. IC_{2d} acts as a buffer to drive the small speaker, LS_1. When the key is inserted, reed switch RS_9 closes and provides power to the circuit. At the same time, relay RL_A energises and latches the power supply on via the contacts RL_{A1}. When RS_9 is opened by removing the correct (or incorrect) key, the circuit continues to function, i.e. the alarm to sound or the solenoid to be energised. Push switch, SW_1, acts as a reset to disable the circuit.

The eight reed switches, RS_1 to RS_8, provide 512 possible combinations. The key is designed using small permanent magnets. The reed switches are wired so that the magnets have to close some reed switches but leave others open. If a magnet closes a reed switch which has to be left open, the alarm sounds since an input to the gate goes low.

Further ideas

(a) Use on/off switches instead of reed switches.
(b) Use CMOS NAND gates instead of TTL. The 4068 CMOS device is an 8-input NAND gate; the 4011 is a quad 2-input NAND gate.

17.3 Sequential Lock

Instead of a solenoid being activated by the correct combination of inputs to a combinational logic circuit, you could design a sequential lock which requires the correct sequence of input pulses to be delivered to a counting circuit.

There are a number of counters which could be used. Examine the possibility of using two or more CMOS 4017 decade counters (Section 6.2) in the design of a sequential lock. The diagram in figure 17.2 is a guide to one possible solution.

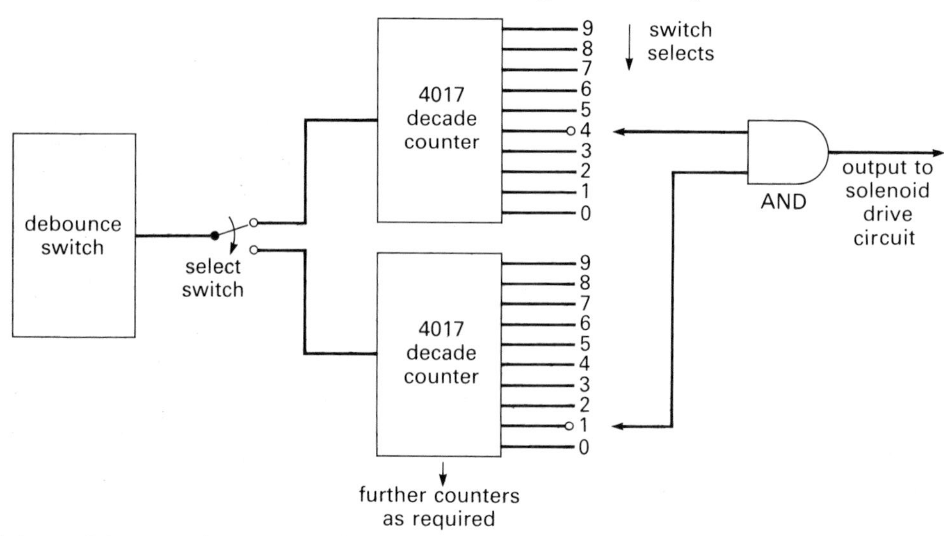

Figure 17.2 Possible system for a sequential lock

Each 4017 is clocked in turn using the debounced switch (e.g. a Schmitt trigger — see Chapter 9). After the correct number of pulses has been input to one 4017 counter, the pulses are directed to the next 4017 counter. When the correct set of pulses has been delivered to the counters, the AND gate has received all highs at its inputs; its output goes high and activates a solenoid through a driver transistor. How would you use rotary switches at the output of each counter so that you could change the sequence of pulses required to energise the solenoid?

Further ideas

(a) Alternatively the TTL 7493 or the CMOS 4516 4-bit counters (Section 5.5) could be used in a sequential lock. The correct 4-bit word is directed from each counter in turn and passed to four inputs of an 8-input NAND gate. An inverter between one or more lines from the counters and the NAND gate would allow the code to be selected as, say, 0110; i.e. inverters are placed on the A and D lines so the NAND gate receives the code 1111_2 after 6 pulses. Similarly, the second counter could send 1111_2 to the other four inputs of the NAND gate when it outputs 0010_2 (after 2 pulses); i.e. inverters are placed on the A, C and D lines. When all inputs to the NAND gate are high, its output goes low; an inverter provides a high to operate a transistor driving a solenoid.

(b) Obviously sequential locks are in demand since the 7225 integrated circuit is dedicated to solving the design of sequential locks. It is worth studying the data sheet for this device.

17.4 LED Scope

Figure 17.3 shows how an array of 100 LEDs can be used to display signals in the audio frequency range. The signals to be displayed are fed to the input (pin 5)

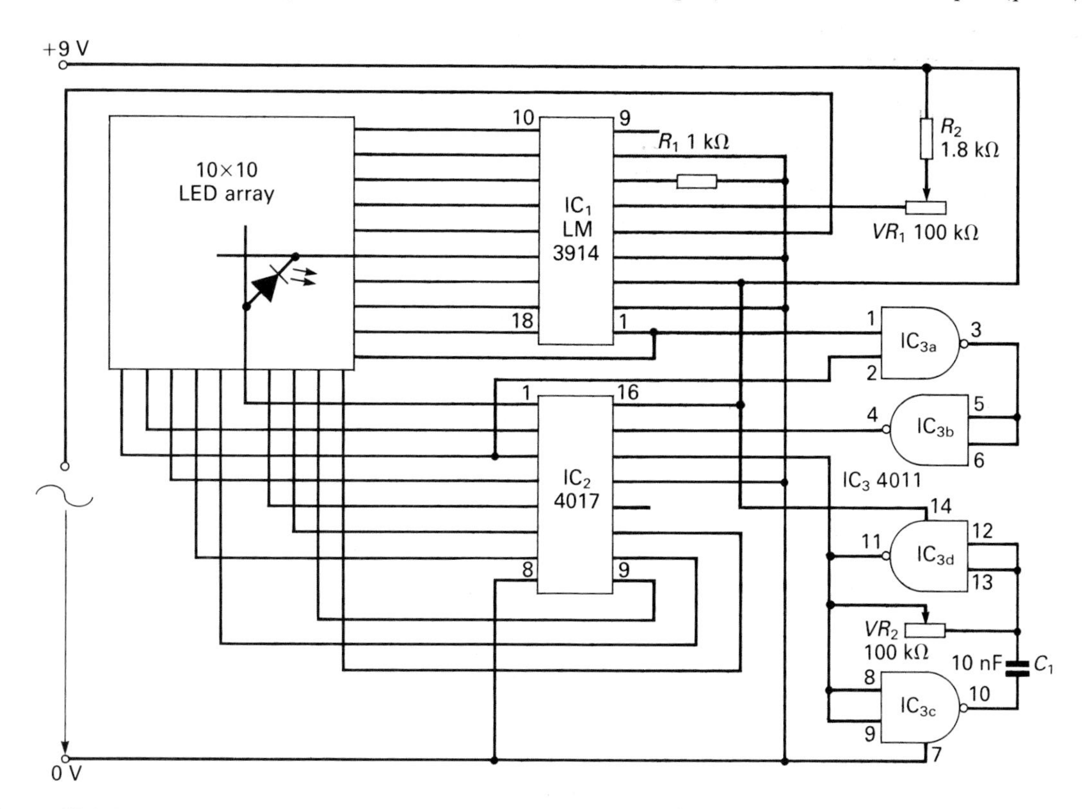

Figure 17.3 LED scope

of a dot/bar driver, IC_1. This device is used to display the strength of an analogue signal, such as a hi-fi VU meter, which shows as a bar of illuminated LEDs, or as a moving dot when the position of an illuminated LED indicates the strength of the signal.

In the circuit shown, IC_1 provides an active low signal to switch on one of ten LEDs in a vertical column. At the same time, the 4017 decade counter scans the vertical rows one after the other. Thus if a steady d.c. signal is applied to the input of IC_1, a horizontal row of LEDs would be lit. A varying input signal, such as a sine wave, is traced out on the LED array. The gates, IC_{3a} and IC_{3b}, provide automatic triggering of the display by ensuring that the 4017 is triggered each time the input signal passes through 0 V. The frequency of the astable based on gates IC_{3a} and IC_{3b} is controlled by VR_2 and C_1 to form the horizontal sweep circuit.

Resistor R_1 controls the current drawn by the LEDs. Pin 7 of IC_1 is connected to an internal voltage reference source so that the current through R_1 is approximately equal to one tenth of the LED current. Thus with R_1 equal to 1.2 kΩ, the LED current becomes equal to about 10 mA.

Book D, Section 4.4 explains the operation of the 3914 dot/bar driver, and Section 6.2 in this book, explains the 4017 decade counter.

17.5 Tachometer

The circuit shown in figure 17.4 uses a purpose-made *tachometer chip*, IC_1, to measure the frequency of rotation of something, like a bicycle wheel, motor shaft or anemometer shaft. IC_1 is actually a frequency-to-voltage converter which means that it produces an output voltage which is directly proportional to the input frequency. Thus the circuit responds to a digital input and produces an analogue output.

Magnetic or optical sensors are generally used to sense the rotational speed of an object without the sensor actually making contact with the object. The two optical sensors shown in figure 17.4 are commonly used to detect the rotation of the object. Each device comprises an infrared LED and a phototransistor sensor mounted in a plastic housing. For the slotted opto-switch, a rotating disc with holes or slots round its circumference rotates in the gap between the LED and phototransistor. In the reflective opto-switch, the phototransistor responds to infrared radiation reflected from, for example, a piece of aluminium foil attached to the rotating object. Each device produces a series of pulses which is fed to IC_1 for conversion to a d.c. voltage. Note that the optimum sensor/reflector distance for the reflective opto-switch is about 5 mm.

The pulses generated by the basic circuit for operating each type of opto-switch are fed to pin 1 of IC_1 via a coupling capacitor, C_1. The circuit is designed to measure frequency in two ranges: 0–500 Hz and 0–5000 Hz, dependent on whether C_3 (5000 Hz range) or C_4 (500 Hz range) are switched in by SW_1. The potentiometer, VR_1, allows the output voltage to be adjusted for calibration of the tachometer. LED_1 simply provides an indication that the tachometer is switched on. Because the current drain of this circuit is so small, the tachometer will operate from a PP9 battery for a long period of continuous use.

Note that it is convenient to mount the 330 Ω and 15 kΩ resistors close to the opto-switch in an assembly which plugs into the circuit. Use an audio signal generator to calibrate the circuit, and a 5 V analogue or digital voltmeter at the output. Switch the circuit to its 500 Hz range. Set the audio frequency generator to 500 Hz and adjust VR_1 so that the output reads 5 V.

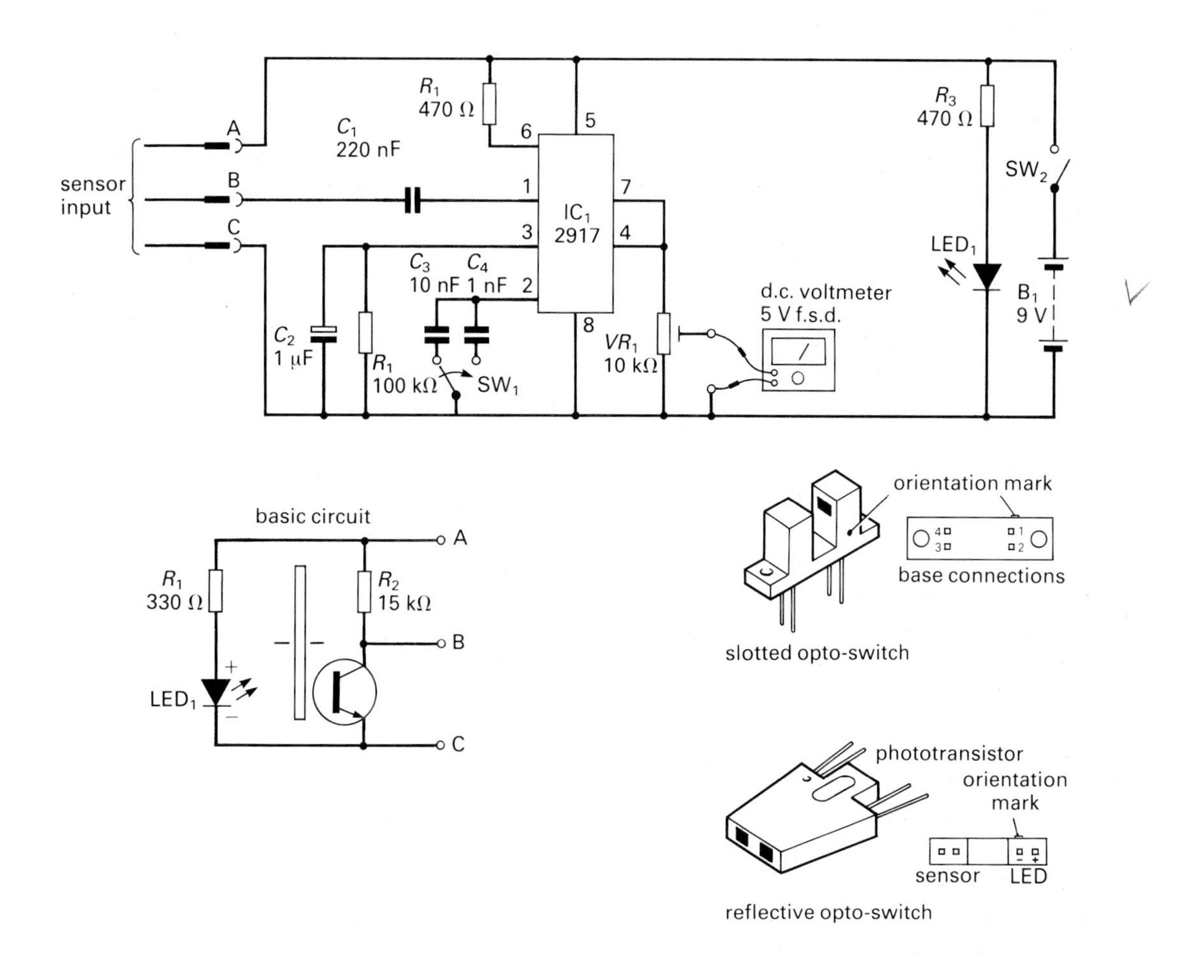

Figure 17.4 Tachometer using the 2917 integrated circuit

17.6 Capacitance Meter

The circuit shown in figure 17.5(a) uses two 555 timers (Book D, Chapter 9) in a dual 556 package, IC_1, to measure the value of a capacitor, C_T, in the range 1 nF to 10 μF in five ranges. The block diagram in figure 17.5(b) shows the way the 555 timers are used. IC_{1a} is wired as an astable which generates pulses at a frequency of about 30 Hz. These pulses trigger the second 555 timer, IC_{1b}, wired as a monostable. The width of the pulses generated by the monostable depends on the value of the test capacitor, C_T, in its timing circuit. Thus a series of constant height and constant width pulses are fed to the meter and provide a reading which is proportional to the value of the capacitor. If C_T is changed for a capacitor of half the value, for a given switch position, the monostable produces pulses of the same frequency but of half the width. The meter averages these pulses to produce a reading which is 50% less. Thus the readings on each range are linear. The variable resistors, VR_1 to VR_6, are used to calibrate the five ranges of the meter.

The value of a capacitor is read on a 1 V f.s.d. digital or analogue voltmeter connected between pin 9 and 0 V of IC_1. The electrolytic capacitor, C_3, smooths the

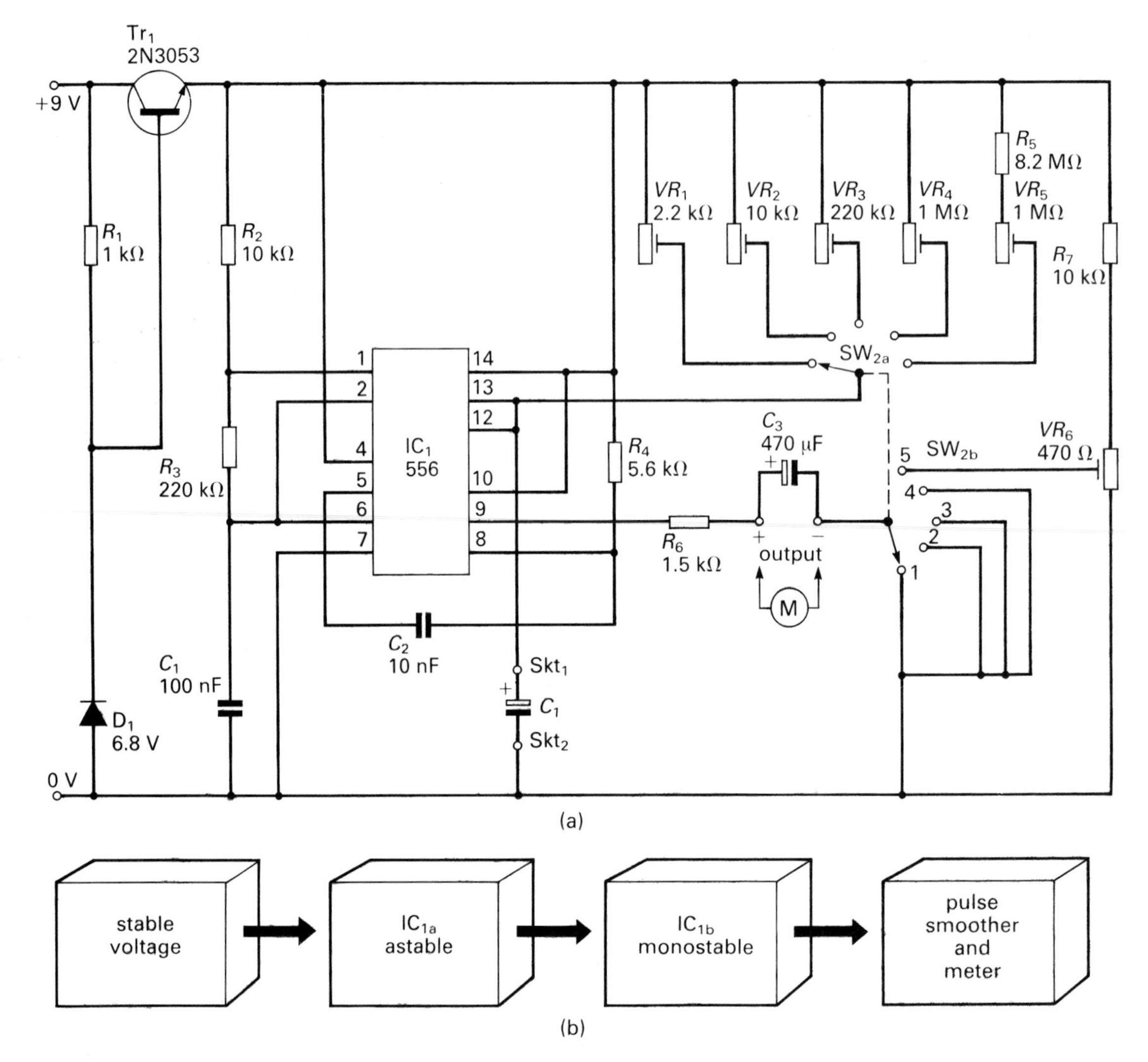

Figure 17.5 Capacitance meter (a) circuit diagram; (b) block diagram

pulses generated by the second 555 timer. On the 1 nF range, VR_6 is required to offset the zero error on the meter due to stray capacitance.

Each range is calibrated by connecting a capacitor of known value across sockets, SKT_1 and SKT_2. Thus for the 1 μF range, a suitable capacitor would be a 0.47 μF (which might actually have a value of 0.52 μF). The range switch, SW_2, is adjusted to the 1 μF range and VR_2 adjusted so the voltmeter reads 0.52 V. The rest of the ranges are calibrated like this by selecting a suitable known capacitor value, for instance the 68 nF on the 100 nF range, except for the lower range of 1 nF.

On the lower range, VR_6 is first adjusted to zero the meter when there is no calibration capacitor across the sockets. Then VR_5 is adjusted with the calibration capacitor in the sockets so that the voltmeter reads the value of the capacitor. Transistor, Tr_1, and Zener diode D_1 provide a cheap way of obtaining a stabilised supply voltage of about 6 V for the circuit. Note that SW_2 should be a 2-pole, 6-way rotary switch, one way not being used. How would you use the extra 'way' on the switch to switch off the power supply?

17.7 Pelican Crossing

The circuit shown in figure 17.6 is designed to sequence the red, green and amber lights at a pelican crossing and to sound a 'peep-peep . .' alarm when the red light is on. The basis of the circuit is a 4017 decade counter, IC_1, (Chapter 6), and two 40106 hex Schmitt triggers, IC_2 and IC_3. IC_{2d} is wired as an astable with a period of about 10 seconds. These pulses are fed into the CLK input of IC_1 so that pins 3, 4 and 7 go high in turn; at the next count, pin 10 goes high and the counter is reset to zero via the connection to pin 15.

As pin 3 goes high, the red LED, D_1, lights; next the green LED, D_2, lights followed by the amber LED, D_3. The amber LED is flashed on and off rapidly by Tr_1, since its collector is connected to the astable based on IC_{2e} and IC_{2f}. When the red LED is on, transistor Tr_2 is switched on and off as its collector is also connected to the 'flashing amber' oscillator. Tr_2 activates the audio oscillator based on IC_{3f} and a 'peep, peep . . .' sound is heard from the piezo-sounder.

17.8 Data Recorder

Figure 17.7 is a systems diagram of a circuit based on CMOS ICs for recording data accumulated in ten consecutive periods. The periods may be adjusted from a fraction of a second to a day or more. Thus rainfall can be measured on ten separate 24-hour days by counting pulses from a microswitch attached to a bucket-type rain gauge. Alternatively, the number of times an anemometer head rotates in a pre-determined period would record average wind speed over ten equal periods.

The pulses from, say, a rain gauge are

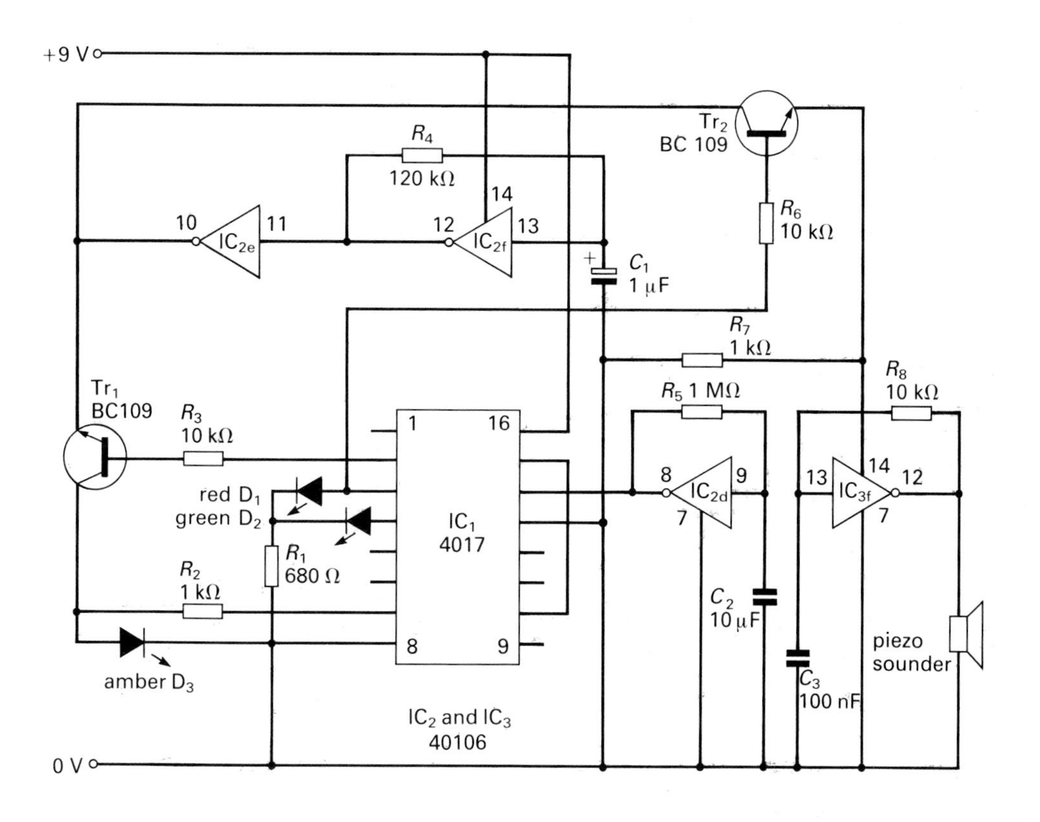

Figure 17.6 Pelican crossing

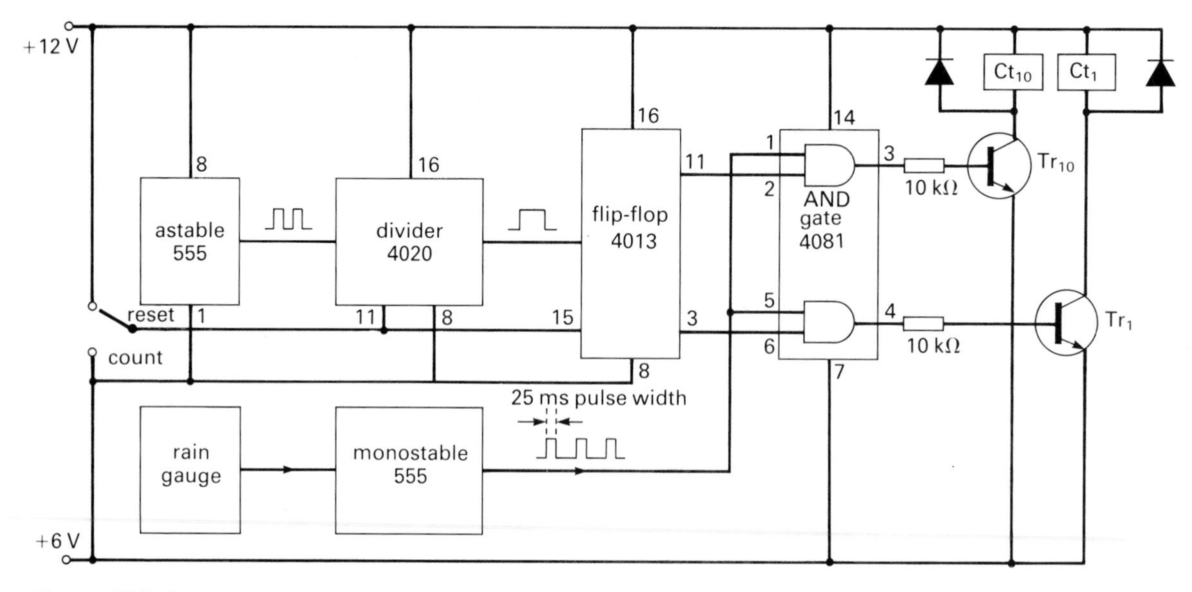

Figure 17.7 Data recorder

'debounced' by a Schmitt trigger (Section 9.3) and 'stretched' by a monostable (Section 9.7). These constant width and height pulses are fed to one of the two inputs of each of ten AND gates made up from three CMOS 4081 ICs. The second input of each of the AND gates is fed from one of the ten outputs of the 4017 decade counter. Since only one of these ten outputs is high at any time, only one AND gate at a time passes pulses from the monostable to the base of a transistor which operates the counter, e.g. Ct_1. There are ten of these counters, Ct_1 to Ct_{10}, which could be electromagnetic counters. Thus if the 5th output from the 4017 is high, for instance during the 5th hour or 5th day, any pulses produced by the rain gauge are recorded on the 5th day counter.

The pulses which clock the 4017 and determine the sampling period are derived from pulses generated by the 555 timer wired as an astable. These pulses are divided by the 14-stage binary counter, the 4020. The 4020 divides the period of an input pulse by a maximum factor of 16 384 (Section 6.2). Thus if the period of the 555 astable is adjusted to 5.3 s, the output pulse from the 4020 has a period of $5.3 \times 16\,384 \simeq 86\,400$ seconds, i.e. 24 hours. Other counting periods can be easily accommodated by the system. The 4020 and 4017 may be reset at the start of a recording session.

Further ideas

(a) The sampling periods may be more precisely set by using a crystal oscillator (Section 9.8) instead of the 555 timer.
(b) The counters may be decade counters, e.g. 4017s or seven-segment displays.

17.9 Music Box

In the application shown in figure 17.8, a CMOS 4051 1-to-8 analogue switch, IC_3, is used to switch one of the resistors R_4 to R_{11}, into the timing circuit of the 555 timer, IC_4, wired up as an astable. These resistors are selected using a binary select code, CBA, at the input of the 4051. If this code is, say, binary 110 (decimal 6), the sixth output (pin 2) of the 4051 goes low and resistor R_{10} is switched into the timing circuit of IC_4. The 4051 is described in Section 8.5, the 555 timer in Book D, Chapter 9, and the BCD counter in Section 5.5.

The rate at which the binary code presented to the 4051 changes is

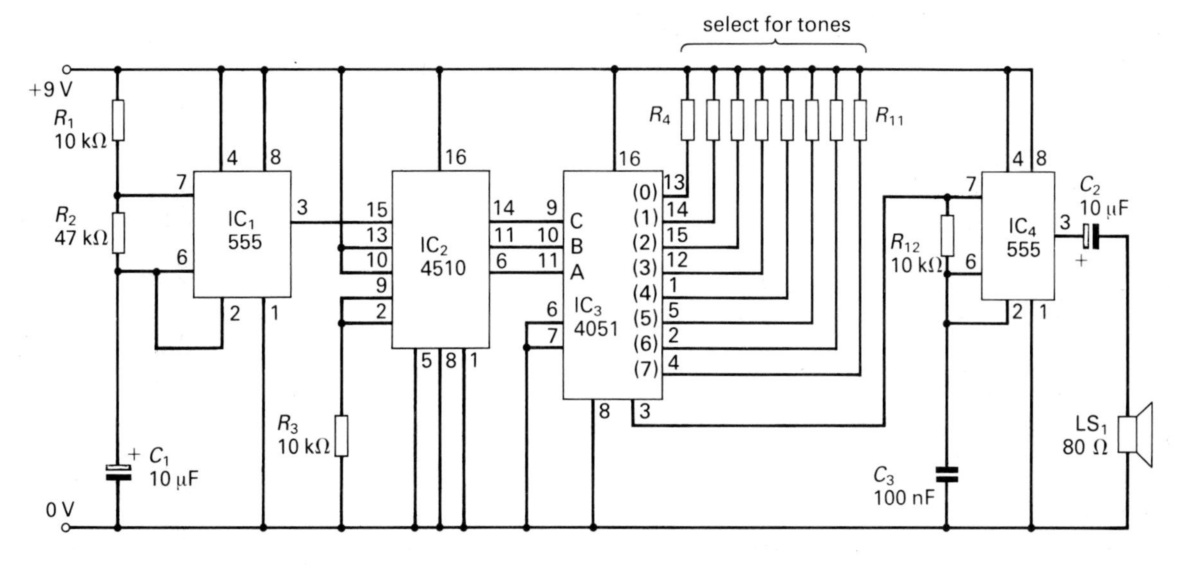

Figure 17.8 Music box

determined by a second astable, IC_1. This feeds pulses at a frequency of about 1 Hz to the clock input, pin 15, of the BCD counter, IC_2. This counter is wired up to reset to 0 when its binary output reaches 1000 by means of the connection between pin 9 (the D output) and the RESET, pin 2. Thus the circuit generates an eight-note tune, its notes determined by the values of R_4 to R_{11}.

The circuit offers a lot of scope for experiment. The value of C_1 changes the rate at which the eight-note tune is played. The value of C_3 determines the key in which the tune is played. A long note can be introduced by making two adjacent resistors have the same value.

17.10 Bat Detector

Over millions of years bats have developed a superbly sensitive sonar system which is exquisitely adapted for navigation and finding food while flying about in the dark. Their built-in sonar produces bursts of ultrasonic waves which bounce back from objects, including flying insects, to give bats a 'picture' of their surroundings in probably as much detail as our sight gives us.

As a matter of interest, the use of echolocation by bats wasn't known until about 1940. Some scientists couldn't believe what they heard. In 1940, sonar and radar were still classified military secrets and the idea that bats had achieved something akin to the latest triumphs of electronic engineering struck them as being rather fanciful daydreaming.

Figure 17.9 A bat detector

So much for open-minded science!

The circuit shown in figure 17.10 is hardly a triumph of electronic engineering, but it let's you in on the bats 'silent' navigation system. You might like to think of the circuit as an electronic 'translator' that enables you to eavesdrop on a little bit of 'batspeak'. The ultrasonic detector, US_1, picks up the bursts of ultrasonic sound bats make. These bursts are then greatly amplified by two audio amplifiers, ICIa and ICIb, connected in series. IC1 is a dual preamplifier normally used for audio stereo applications. The amplified ultrasonic frequencies are then fed to a frequency divider, IC2, which reduces the inaudible sounds to a series of 'chick-chick-chick....' sounds in the headphones connected to one of the outputs of the frequency divider. You may select an output which gives a satisfactory pitch to each 'chick'.

The repetition rate of these bursts varies from about 20 to 200 per second depending on the species of bat and what it is doing. For example, when a Little Brown Bat is cruising around looking for an insect, it produces bursts of ultrasonic sound at around 10 per second. But when it detects an insect and starts to move in on an interception course, the frequency of these bursts goes up to 200 per second which sounds like a 'buzz', or a 'burp' through the earphones! The frequency of the ultrasonic component of each burst of sound is not a fixed but varies from, say, 100 kHz to about 20 kHz. Because of the limited frequency response of the ultrasonic detector, the circuit only amplifies a small part of this frequency spread, but you'll hear enough 'batspeak' to make you want to know more about these fascinating animals. Figure 17.9 shows a bat detector assembled as portable battery-operated device.

CAUTION

When studying bats, you should take every care not to disturb their roots or interfere in any way with their private lives. The fascinating habits and the highly evolved navigation system of these mammals is described in a number of excellent books including:

Richardson P (1985), *Bats*, London Whittet Books

Schober W (1984), *The Lives of Bats,* London Croom Helm

Stebbings R E (1986), *Which Bat Is It?* The Mammal Society

Dawkins R (1986), *The Blind Watchmaker,* Penguin Books

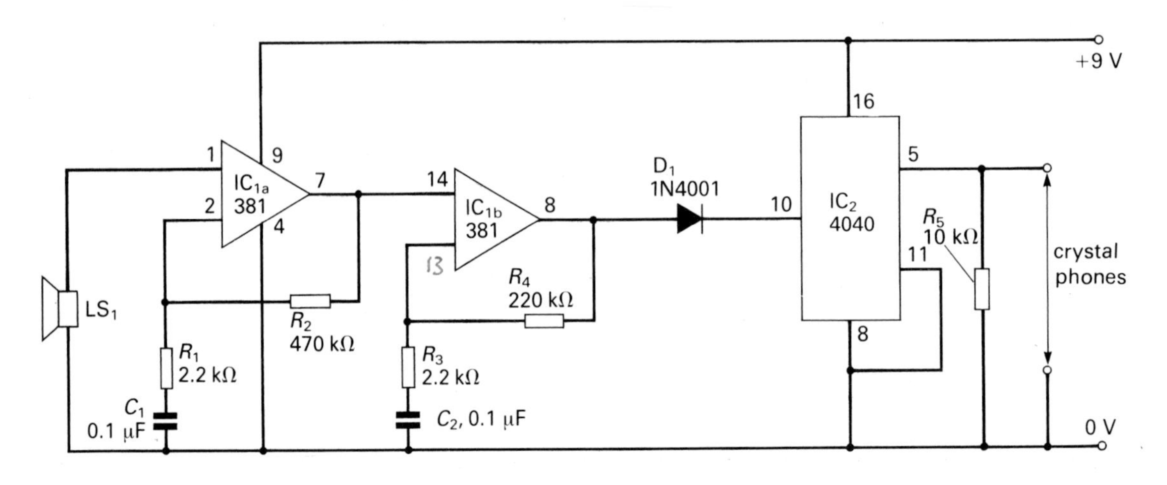

Figure 17.10

17.11 Intruder Alarm

The circuit shown in figure 17.11 has a low standby current yet produces a loud pulse-tone audio alarm when activated by the opening of the reed switch RS_1. All four gates in a CMOS quad 2-input NOR package are used. Gates IC_{1a} and IC_{1b} operate as a bistable. Components C_1 and R_3 ensure that the system is switched to standby when the power switch SW_1 is closed. SW_1 is also used to reset the alarm. The output of this bistable controls an audio oscillator based on gates IC_{1c} and IC_{1d}.

The frequency of the oscillator is determined by components C_2 and R_5. If you use normally-open reed switched, use two permanent magnets one adjacent to the reed switch and one in the door or window to ensure that the switch is open when the door is closed. A number of reed switches should be connected in parallel to monitor the opening of a number of doors and windows. The VMOS transistor, Tr_1, interfaces directly with the CMOS gates to provide a loud sound output from the loudspeaker.

17.12 Auto-turn-off Alarm

The practical circuit shown in figure 17.12 generates an audio frequency alarm which lasts for 10 seconds when SW_1 is momentarily closed. It is based on all four NOR gates in a 4001 package. IC_{1a} and IC_{1b} are connected as a monostable which gates the astable based on IC_{1c} and IC_{1d}. The time delay is determined by R_1 and C_1 and the audio frequency by R_3 and R_4. The VMOS transistor directly interfaces with the CMOS device and a surprisingly loud note is produced by the loudspeaker.

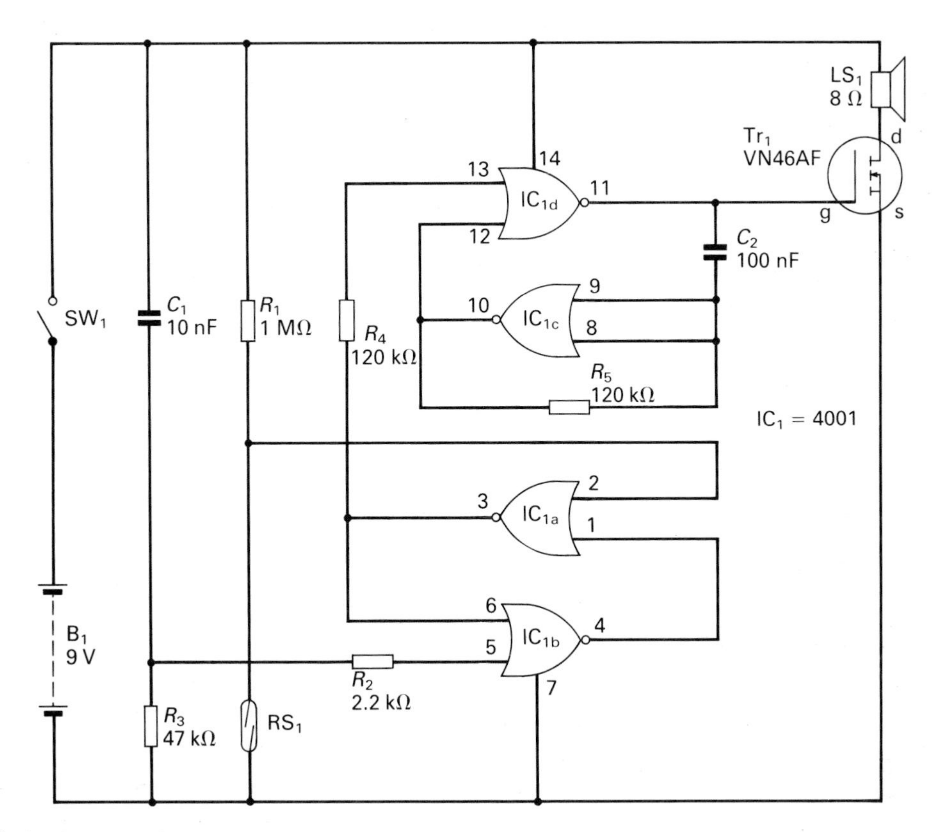

Figure 17.11 Intruder alarm

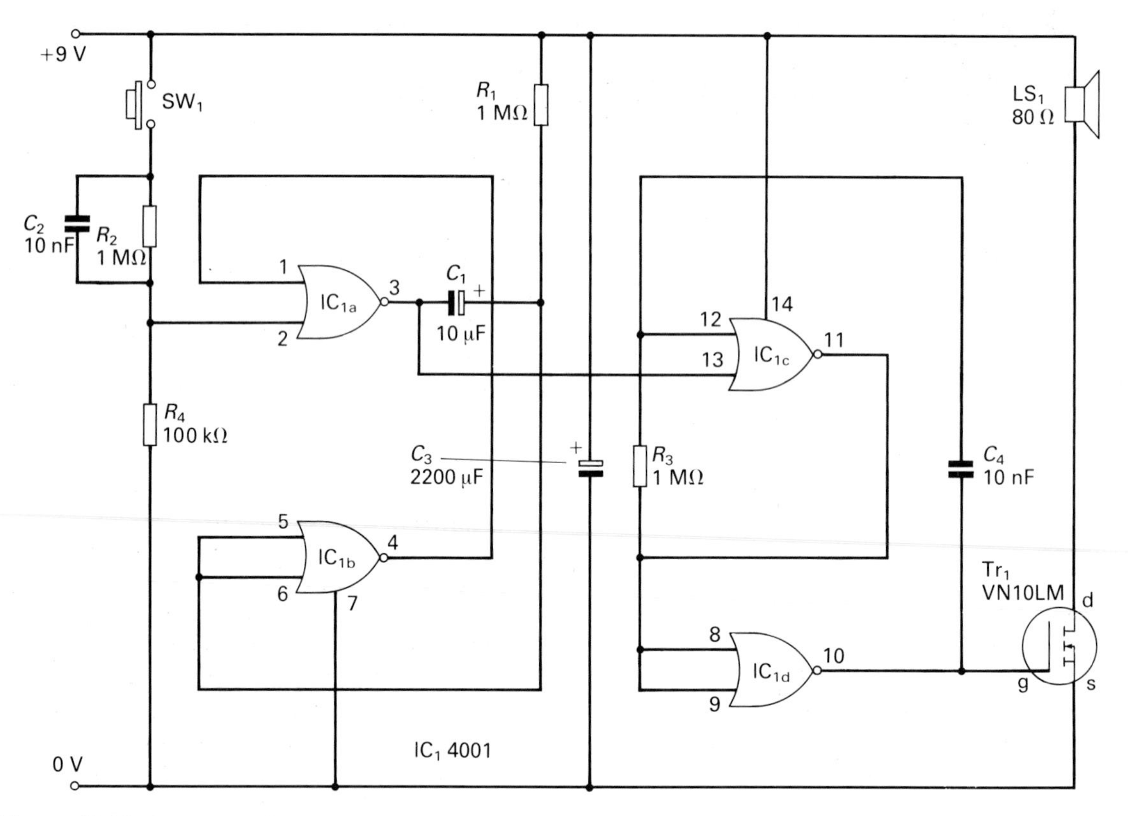

Figure 17.12 Auto turn-off alarm

18 Project Modules

18.1 What they are

At the end of each book of *Basic Electronics* there are a number of practical projects for you to build. These projects are called Project Modules and there are thirty five of them in all. This chapter explains how to build the seven Project Modules shown in figure 18.1. They are:

E1: Logic Gates
E2: BCD Counter
E3: Keyboard Encoder
E4: 4-bit Magnitude Comparator
E5: BCD-to-Decimal Decoder
E6: Infrared Remote Control (two projects, a receiver and a transmitter)
E7: 64-bit Memory

The Project Modules enable you to build up a set of electronic building blocks which can be connected together in various ways to design useful and interesting electronic systems. Details are provided for assembling each Project Module on a printed circuit board (PCB), and for interconnecting it with other Project Modules using flying leads.

Before assembling the circuits, you should read Book A, Section 6.3 which gives guidance on the preparation of the PCBs. You should also read Book A, Section 12.2 which gives guidance on the CMOS devices used in all seven Project Modules, The references to other parts of *Basic Electronics* provide further information on the devices and circuits used in the Project Modules. The examples for using the Project Modules to design electronic systems should enable you to solve other problems which wholly or in part have an electronic solution.

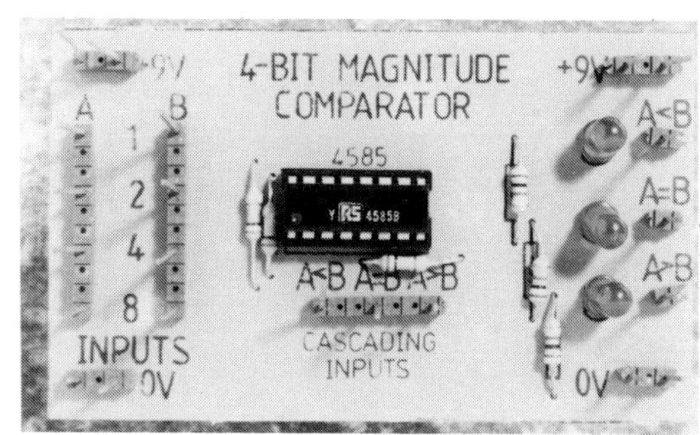

Figure 18.1 Project Modules E1, E2, E3 and E4

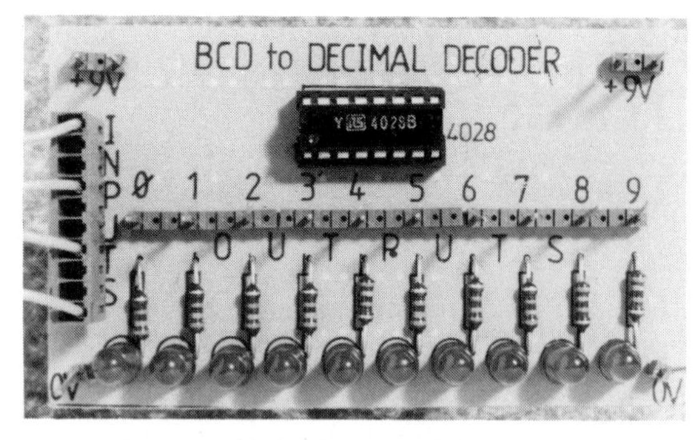

Figure 18.1 Project Modules E5, E6 and E7

18.2 *Project Module* E1

Logic Gates

What it does

This module provides the basic logic functions of AND, OR, AND, and NAND by plugging the appropriate CMOS integrated circuit into the 14-way IC socket. This flexible use of the same basic Project Module is possible since the input and output pins of the gates in the four CMOS ICs used are identical. Logic gates can be used in a variety of control and decision-making applications as explained in Chapters 3, 4 and 15. A NAND gate used as a NOT gate is explained in Chapter 9 as part of the 4-bit memory (Project Module E7).

Circuit

Figure 18.2 shows the basic circuit for all

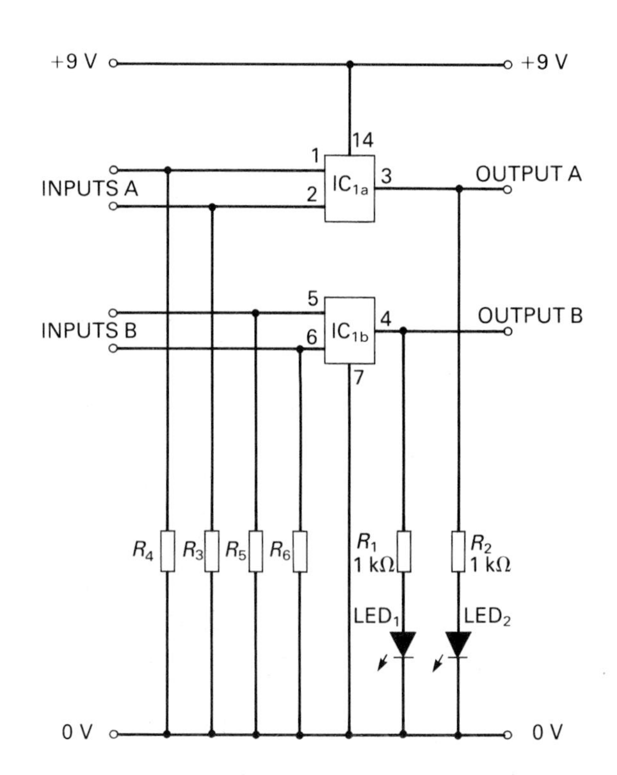

Figure 18.2 Circuit diagram of the Logic Gates

four uses of Logic Gates. Each of the four CMOS integrated circuits to be used with this Project Module contains four 2-input gates, only two of which are used, IC_{1a} and IC_{1b}. The logic state, high or low (or binary 1 or 0, respectively) of the output of each gate is indicated by the LEDs. The two inputs to each gate are held at logic 0 by means of resistors R_3 to R_6. Thus a logic 1 can be applied to a gate by connecting the input to the positive terminal of the power supply.

Components and materials

IC_1: quad 2-input CMOS logic gates as follows: 4001 (NOR gates); 4011 (NAND gates); 4071 (OR gates); 4081 (AND gates)
IC holder: 14-way
R_1 to R_5: fixed-value resistors, values 100 kΩ each, 0.25 W ± 5%
LED_1, LED_2: red light emitting diodes
battery: PP9
battery clip: PP9 type
wire: multistrand, e.g. 7 × 0.2 mm
PCB: 90 mm × 50 mm
connectors: PCB header and PCB socket housing; crimp terminals

PCB assembly

Figure 18.3 shows the layout of the components on the PCB, and figure 18.4 the copper track pattern on the other side of the PCB. See Book A, Section 6.3 for guidance on the preparation of the PCB.

Make sure that the two LEDs are connected the right way round. The cathodes of both should be facing the edge of the board. The terminal pins for the connections on the PCB are made by cutting single and double pins from the PCB header. Single sections of the PCB socket housing are used for terminating wire ends for connecting the Logic Gates to the battery and to other Project Modules. Strip 5 mm of insulation from

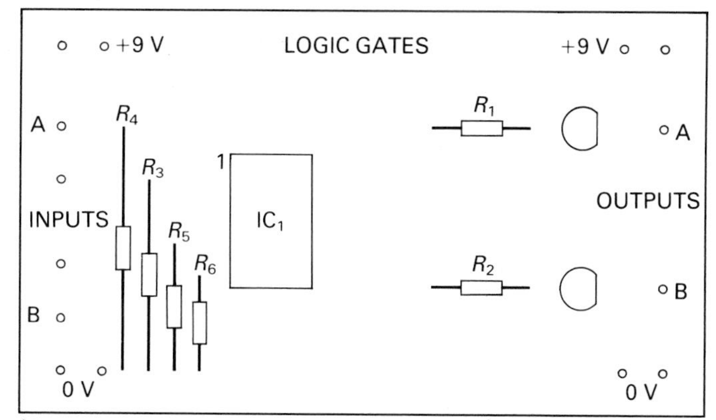

Figure 18.3 Component layout on the PCB (actual size)

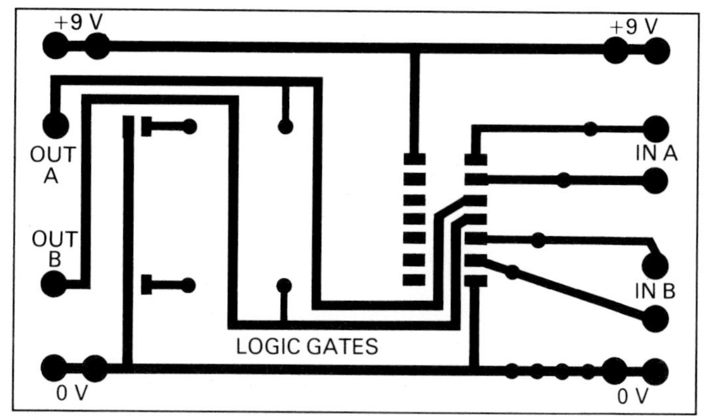

Figure 18.4 Track pattern on the PCB (actual size)

the ends of the wires, and use a crimping tool to squeeze a crimp connector on the bare ends. Push the crimp connector into the PCB socket until it clicks into place.

Testing and use

The best way to test the function of the Logic Gates Module is to refer to the truth tables for the AND, NAND, OR, NOR and NOT gates listed in Sections 3.3 to 3.7. Beginning with the 4081, say, check that the logic gates function as an AND gate. Note that the inputs to each gate are held logic low by the 100 kΩ resistors. Thus on connecting power to the AND gate, the LED showing the output to this gate will be off, i.e. '0 plus 0 equals 0'. Use flying leads to connect the two inputs to + 9 V, i.e. logic 1, and show that when both inputs are at logic 1, the LED is on, i.e. 1 and 1 equals 1. Similarly replace the 4081 by the other devices in turn and check that each IC provides the correct truth table. Note that the NOT gate is obtained by wiring the two inputs of the NAND or NOR gate together as explained in Section 4.4. A NOT gate is used in Project Module E7.

18.3 *Project Module* E2

BCD Counter

What it does

This Project Module counts up or down in binary and displays the 4-bit count on an LED display. It is based on a single integrated circuit, a CMOS 4510 BCD Counter which counts up to binary 1001 (decimal 9). It is possible to replace the 4510 by the 4516 4-bit Counter to count up to decimal 15 (binary 1111). A RESET button is provided so that any count can be reset to zero, i.e. binary 0000. Section 5.5 discusses BCD and 4-bit Counters.

Circuit

Figure 18.5 shows the simplicity of the circuit. The displayed count is reset to zero by momentarily pressing SW_1, or by bringing the reset pin high. This pin enables a reset to be made automatically by a signal from another Project Module. Four further pins enable the count to be connected to other Project Modules, e.g. the 4-bit Magnitude Comparator described in Section 18.5. The up/down pin is taken high if the Counter is to count up, and taken low if it is to count down.

Components and materials

IC_1: BCD Counter, CMOS type 4510; or 4-bit Counter type 4516
IC holder: 16-way
LED_1 to LED_4: red light-emitting diodes
R_1: fixed-value resistor, value 10 kΩ, 0.25 W, ± 5%
R_2 to R_5: fixed-value resistors, values 1 kΩ
SW_1: push-to-make, release-to-break keyboard switch
battery: PP9
battery clip: PP9 type
wire: multistrand, e.g. 7 × 0.2 mm
PCB: 90 mm × 50 mm
connectors: PCB header and PCB socket housing; crimp connectors

PCB assembly

Figure 18.6 shows the layout of the components on the PCB, and figure 18.7 the copper track pattern on the other side of the PCB. See Book A, Section 6.3 for guidance on the preparation of the PCB.

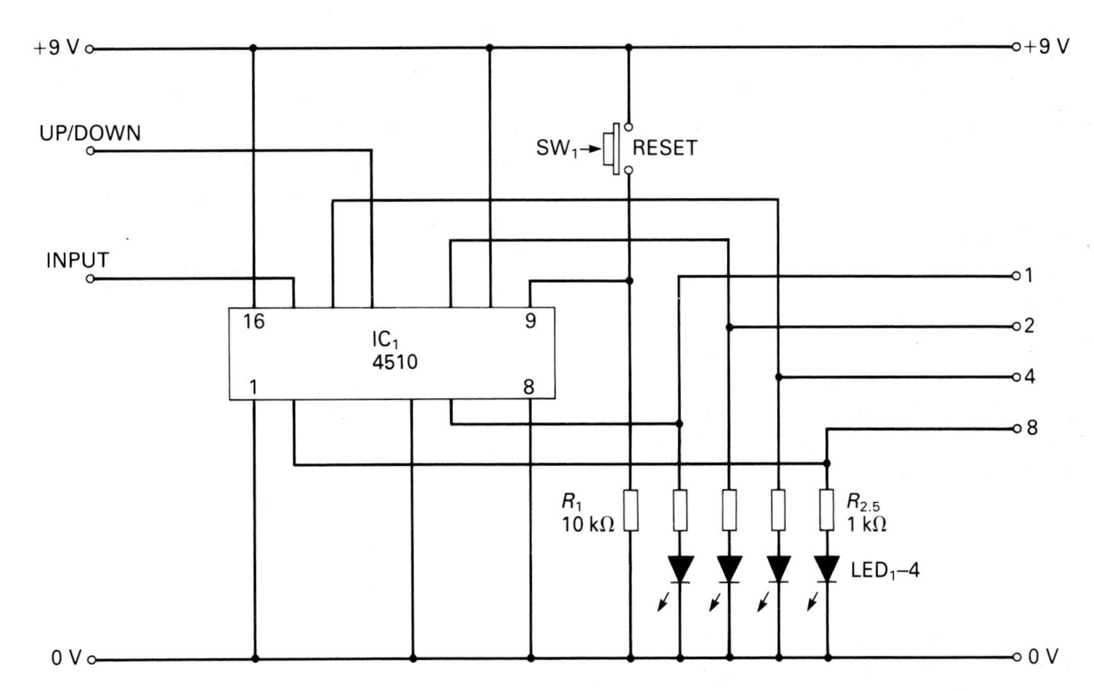

Figure 18.5 Circuit diagram of the BCD Counter

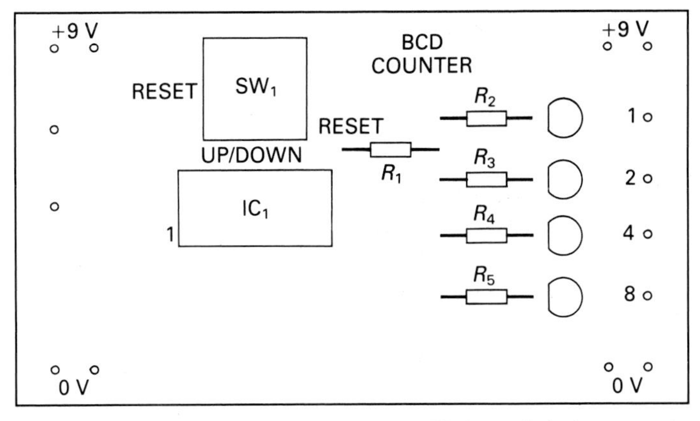

Figure 18.6 Component layout on the PCB (actual size)

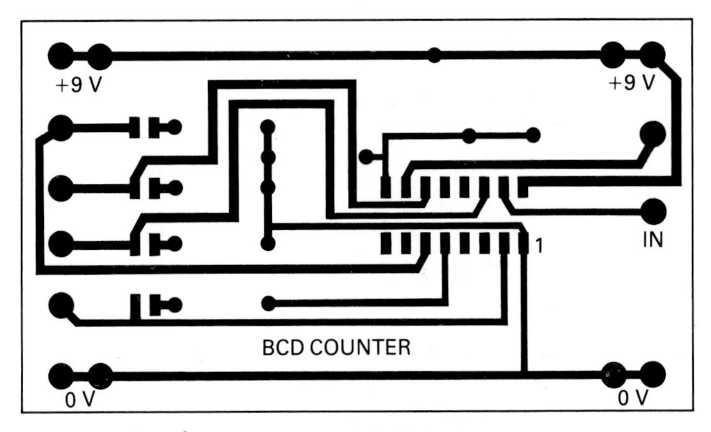

Figure 18.7 Track pattern on the PCB (actual size)

The terminal pins for the connections on the PCB are made by cutting single and double pins from the PCB header. Single sections of the PCB socket housing are used for terminating wire ends for connecting the BCD Counter to the battery and to other Project Modules. Strip 5 mm of insulation from the ends of the wires, and use a crimping tool to squeeze a crimp connector on the bare ends. Push the crimp connector into the PCB socket until it clicks into place.

Testing and use

Wire up the Pulser (Project Module A2) so that it produces pulses at intervals of about 1 s. Connect the output of the pulser to the input of the BCD Counter. Connect together the + 9 V and 0 V power supply pins on the two Modules. Connect the up/down pin on the BCD Counter to + 9 V and momentarily press SW_1 to reset the BCD Counter to binary 0000, i.e. all LEDs off. Connect a PP9 battery to the Pulser and note that the LEDs will show an increasing binary count. Now connect the up/down pin to 0 V and note that the display counts down in binary.

Connect the Relay Driver (Project Module B2) to the output of the BCD Counter so that one relay is energised at 2 s intervals and the other at 8 s intervals.

The BCD Counter will be found useful when inputting data as a binary count to the 4-bit Magnitude Comparator (Project Module E4).

How would you use Logic Gates (Project Module E2) using the CMOS device 4081 (an AND gate) so that the BCD Counter automatically resets after every fifth count?

18.4 Project Module E3

Keyboard Encoder

What it does

This Project Module produces the 4-bit binary equivalent of the value of any one of 16 keys. The binary word is displayed on four LEDs. The outputs from these LEDs provide a preset 4-bit word which can be used in conjunction with the 4-bit Magnitude Comparator (Project Module E4) for control applications.

Circuit

Figure 18.8 shows the simplicity of the circuit which is based on a single integrated circuit, a 74C922. The value of the 4-bit word generated when a key is pressed is displayed on LED_1 to LED_4. The general principles of encoders and decoders are described in Chapter 7.

Components and materials

IC_1: key encoder, CMOS type 74C922
IC holder: 1 × 18-way
LED_1 to LED_4: light emitting diodes
R_1 to R_4: fixed-value resistors, all 1 kΩ, 0.25 W ± 5%
SW_1 to SW_{16}: push-to-make, release-to-break keyboard switches
C_1, C_2: polyester capacitors, values 10 nF and 100 nF, respectively
battery: PP9
battery clip: PP9 type
wire: multistrand, e.g. 7 × 0.2 mm
PCB: 90 mm × 50 mm
connectors: PCB header and PCB socket housing; crimp connectors

PCB assembly

Figure 18.9 shows the layout of the components on the PCB, and figure 18.10

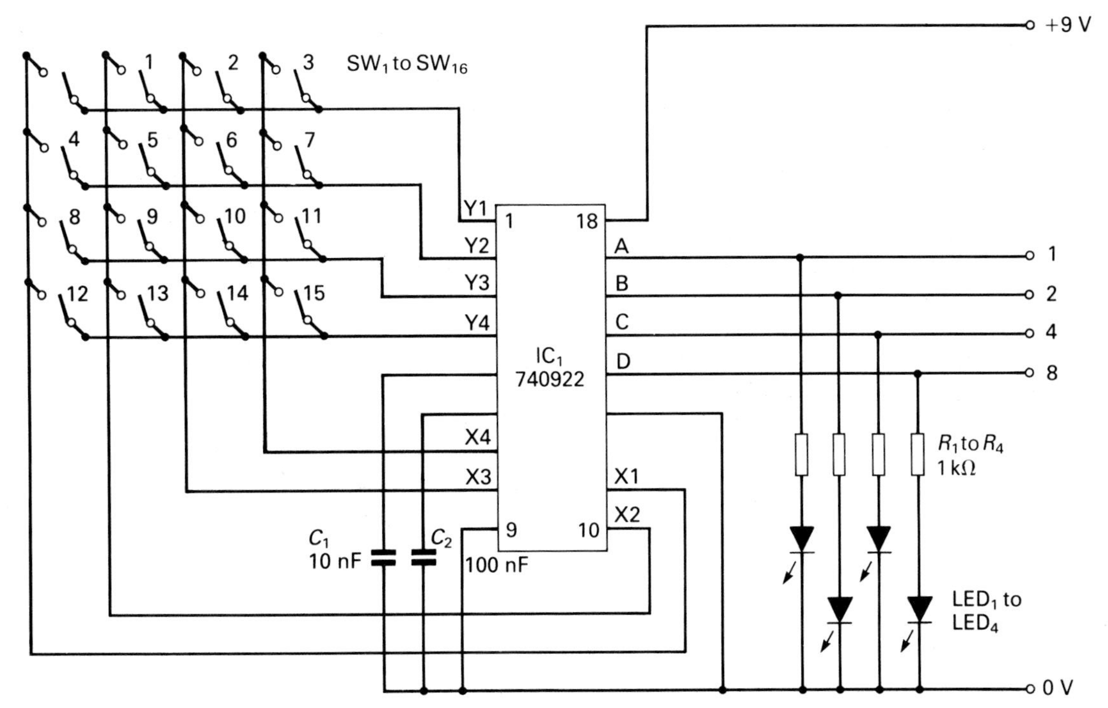

Figure 18.8 Circuit diagram of the Keyboard Encoder

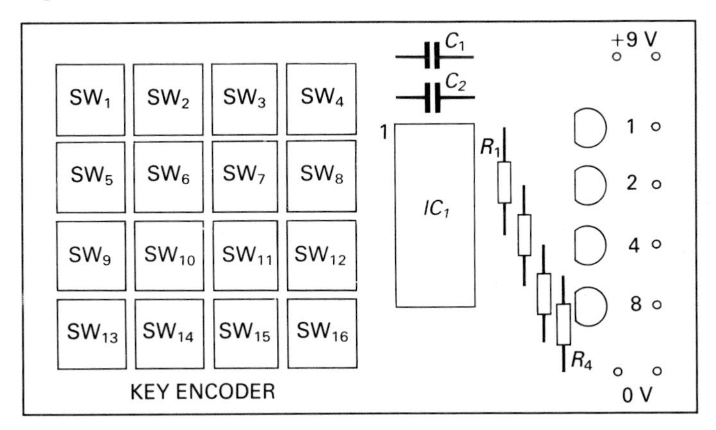

Figure 18.9 Component layout on the PCB (actual size)

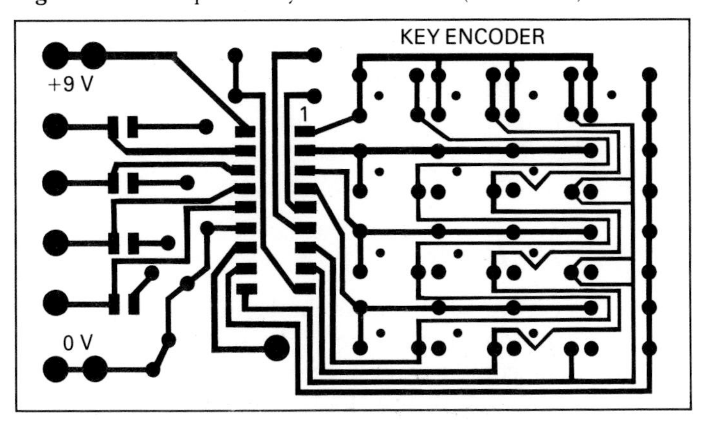

Figure 18.10 Track pattern on the PCB (actual size)

the copper track pattern on the other side of the PCB. See Book A, Section 6.3 for guidance on the preparation of the PCB.

The terminal pins for the connections on the PCB are made by cutting single and double pins from the PCB header. Single sections of the PCB socket housing are used for terminating wire ends for connecting the Keyboard Encoder to the battery and to other Project Modules. Strip 5 mm of insulation from the ends of the wires, and use a crimping tool to squeeze a crimp connector on the bare ends. Push the crimp connector into the PCB socket until it clicks into place.

Testing and use

Connect a PP9 battery to the Keyboard Encoder and press one of the keys SW_1 to SW_{16}. Note that the binary equivalent of the key pressed appears on the LED display. Check that all the keys work. This Project Module enables decimal numbers to be input as binary data to digital systems, e.g. numerically-controlled machinery and security devices. The Keyboard Encoder provides a preset input to the 4-bit Magnitude Comparator (Project Module E4) and the 64-bit Memory (Project Module E7).

18.5 *Project Module* E4

4-bit Magnitude Comparator

What it does

This Project Module shows when one 4-bit word is greater than, equal to or less than a second 4-bit word. It is used with the Keyboard Encoder (Project Module E3) and the Relay Driver (Project Module B2) for control applications.

Circuit

Figure 18.11 shows the 4-bit magnitude comparator is based on a single integrated circuit IC_1. As explained in Section 14.5, one 4-bit word is fed to inputs A and the second 4-bit word to inputs B. If the value of the word on A is greater than the value of the word on B, i.e. $A > B$, LED_1 lights; if $A = B$, LED_2 lights; and if $A < B$, LED_3 lights. The three output signals from these LEDs can be used to operate the Relay Driver (Project Module B2), e.g. when $A = B$ a d.c. motor can be switched on. This circuit is the digital equivalent of a comparator circuit based on an op amp which responds to analogue input voltages, not digital words.

Components and materials

IC_1: 4-bit Magnitude Comparator, CMOS type 4585
IC holder: 1 × 16-way
LED_1 to LED_3: light emitting diodes
R_1 to R_3: fixed-value resistors, all 100 kΩ, 0.25 W, ± 5%
R_4 to R_6: fixed-value resistors, all 1 kΩ, 0.25 W ± 5%
battery: PP9
battery clip: PP9 type
wire: multistrand, e.g. 7 × 0.2 mm; single strand
PCB: 90 mm × 50 mm
connectors: PCB header and PCB socket housing; crimp connectors

PCB assembly

Figure 18.12 shows the layout of the components on the PCB, and figure 18.13 the copper track pattern on the other side of the PCB. See Book A, Section 6.3 for guidance on the preparation of the PCB.

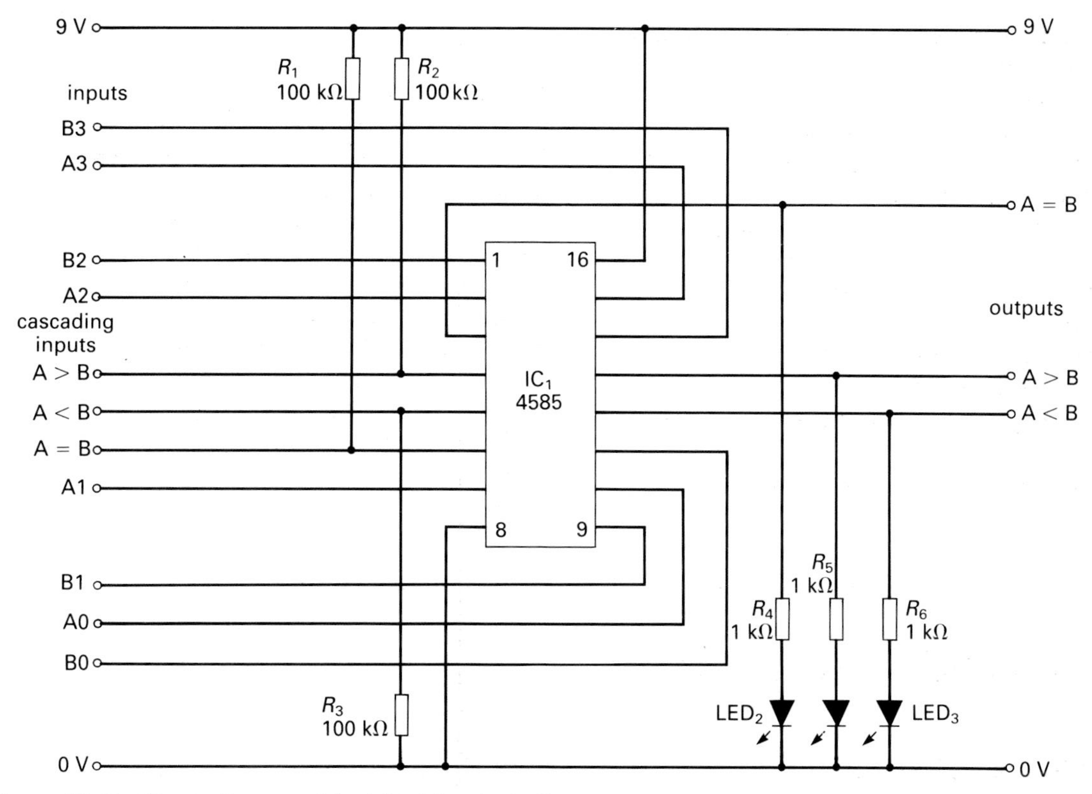

Figure 18.11 Circuit diagram of the 4-bit Magnitude Comparator

The terminal pins for the connections on the PCB are made by cutting single and double pins from the PCB header. Single sections of the PCB socket housing are used for terminating wire ends for connecting the 4-bit Magnitude Comparator to the battery and to other Project Modules. Strip 5 mm of insulation from the ends of the wires, and use a crimping tool to squeeze a crimp connector on the bare ends. Push the crimp connector into the PCB socket until it clicks into place.

Testing and use

Figure 18.14 shows a control system which can be designed using the 4-bit Magnitude Comparator and three other Project Modules.

The Schmitt Trigger (Project Module B1) and the BCD Counter (Project Module E2) provide a 4-bit binary count of the number of times the LDR is covered and uncovered. A maximum binary value of 1111 can be obtained if a 4516 device is used in the BCD Counter. This 4-bit count is fed to inputs A of the 4-bit Magnitude Comparator. The 4-bit binary output from the Keyboard Encoder (Project Module E3) is fed to inputs B of the 4-bit Magnitude Comparator.

Press, say, key number 7 on the Keyboard Encoder to provide a binary value of 0111 on inputs B of the 4-bit Magnitude Comparator. Reset the BCD Counter to zero, and note how the three LEDs on the 4-bit Magnitude Comparator change as the count increases on inputs A. Only one LED is on at any time. As the binary counts increase from 0000, LED$_1$ is on since the binary value on input A is less than that set on input B. When the binary value on inputs A equals the binary value on inputs B, LED$_2$ lights; LED$_3$ lights for the next count.

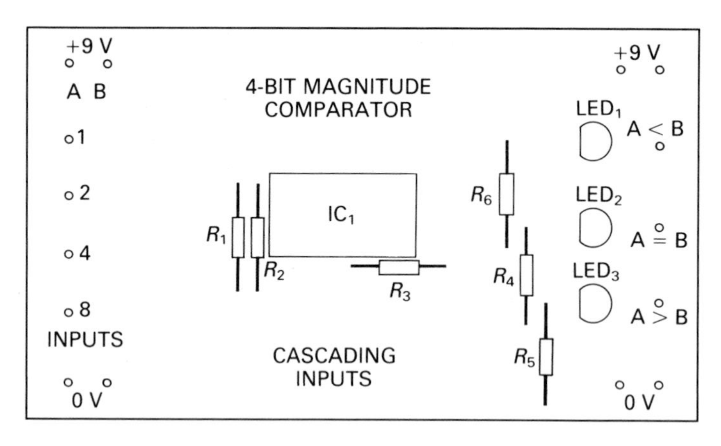

Figure 18.12 Component layout on the PCB (actual size)

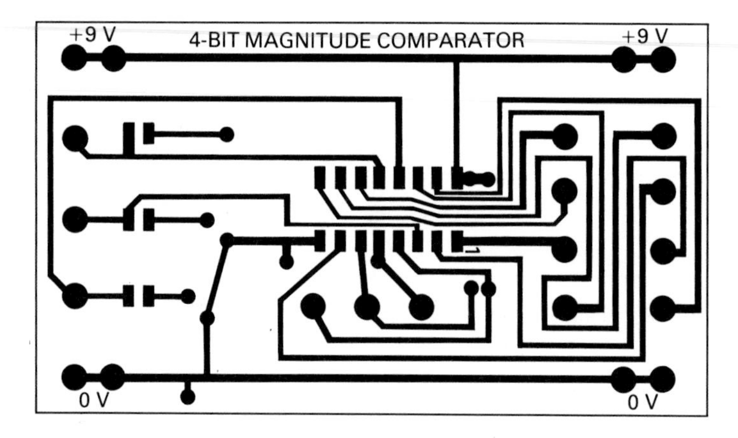

Figure 18.13 Track pattern on the PCB (actual size)

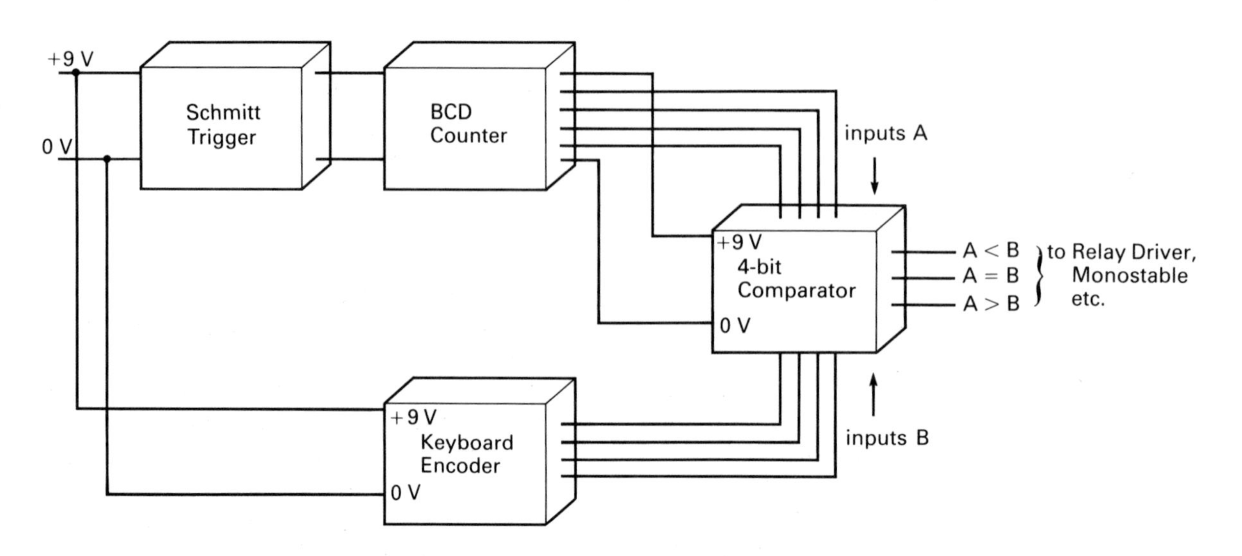

Figure 18.14 System to compare two 4-bit binary words

Questions

1 Connect the Pulser (Project Module A2) and the Piezo-sounder (Project Module A3) to the A = B output so that when the binary count equals that preset on the Keyboard Encoder an audio alarm sounds.
2 How would you use the Monostable (Project Module B4), and the Relay Driver (Project Module B2) so that a d.c. motor is driven for 10 s when the two binary numbers are equal?
3 How would you automatically reset the BCD Counter to 0000 after the 4-bit Magnitude Comparator has shown A = B?
4 Use the Two-Digit Counter (Project Module D7) to give an indication of the number of counts delivered to the 4-bit Magnitude Comparator.
5 Devise a model of a packaging system which will automatically deliver to a loading bay a preset number of objects moving along a production line.
6 Devise a model of a 'bottle filler' which will automatically put a preset number of items, e.g. pills, in bottles passing along a conveyor belt.

18.6 *Project Module* E5

BCD-to-Decimal Decoder

What it does

This Project Module converts a 4-bit word into one of ten decimal outputs. Thus it decodes a binary number into a more familiar decimal value. The decoded binary number is displayed on four LEDs. This Project Module is based on a single integrated circuit, a CMOS device type 4028. Each of the ten outputs from the BCD–Decimal Decoder can be accessed for driving a relay, as suggested for the receiver of the Infrared Remote Control (Project Module E6).

Circuit

Figure 18.15 shows the simplicity of the circuit which operates from a 9 V supply. A BCD number is presented to IC_1 via its A to D inputs. The decimal equivalent of this number is displayed on one of the ten LEDs. Thus if the binary number input to IC_1 is 0111, LED_7 lights. Decoders of this type are described in Section 7.3.

Components and materials

IC_1: BCD–decimal decoder, CMOS type 4028
LED_1 to LED_{10}: red light emitting diodes
IC holder: 1 × 16-way
R_1 to R_{10}: fixed-value resistors, all values 1 kΩ, 0.25 W ± 5%
battery: PP9
battery clip: PP9 type
wire: multistrand, e.g. 7 × 0.2 mm
PCB: 90 mm × 50 mm
connectors: PCB header and PCB socket housing; crimp connectors

PCB assembly

Figure 18.16 shows the layout of the components on the PCB, and figure 18.17 the copper track pattern on the other side of the PCB. See Book A, Section 6.3, for guidance on the preparation of the PCB.

The terminal pins for the connections on the PCB are made by cutting single and double pins from the PCB header. Single sections of the PCB socket housing are used for terminating wire ends for

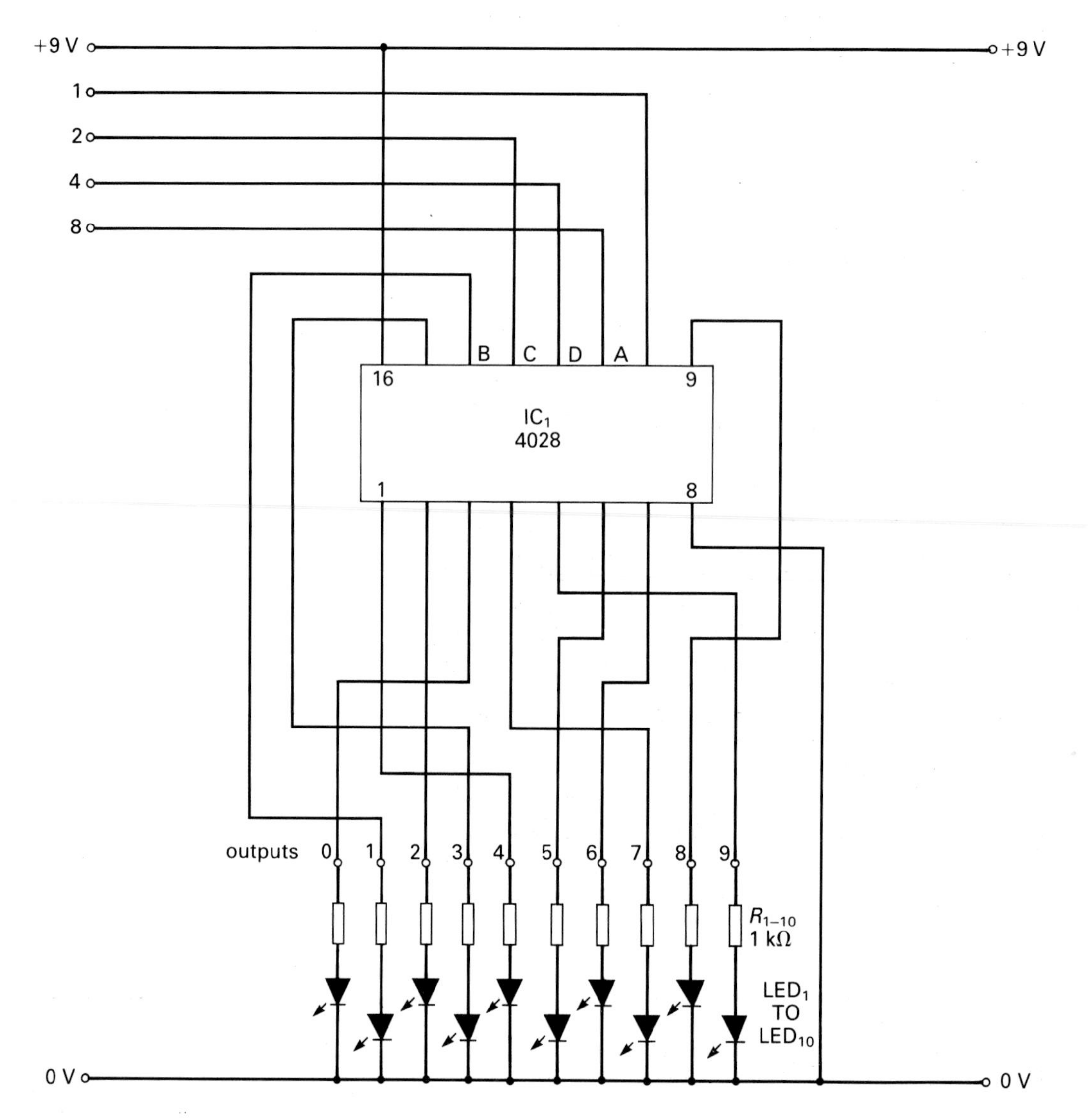

Figure 18.15 Circuit diagram of the BCD to decimal decoder

connecting the BCD–Decimal Decoder to the battery and to other Project Modules. Strip 5 mm of insulation from the ends of the wires, and use a crimping tool to squeeze a crimp connector on the bare ends. Push the crimp connector into the PCB socket until it clicks into place.

Testing and use

The simplest way to test the BCD–Decimal Decoder is to connect it to the BCD Counter (Project Module E2) which is operated by the Pulser (Project Module A2) as explained in Section 18.3. Check that the BCD–Decimal Decoder decodes the binary signal into a decimal value. Thus, you should find that if the binary word presented to the BCD–Decimal Decoder is 1001 (decimal 9), LED_9 on the BCD–Decimal Decoder lights.

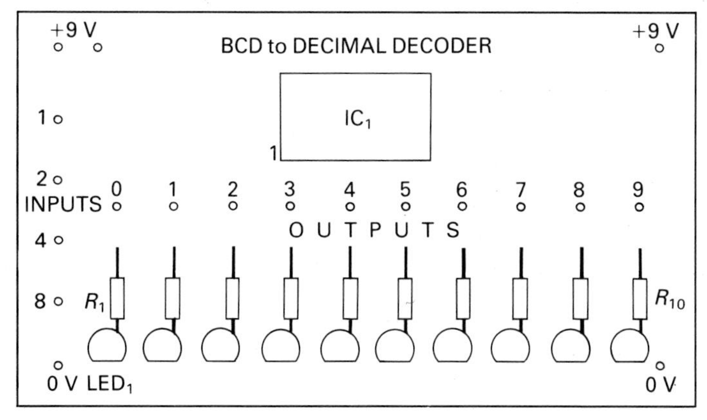

Figure 18.16 Component layout on the PCB (actual size)

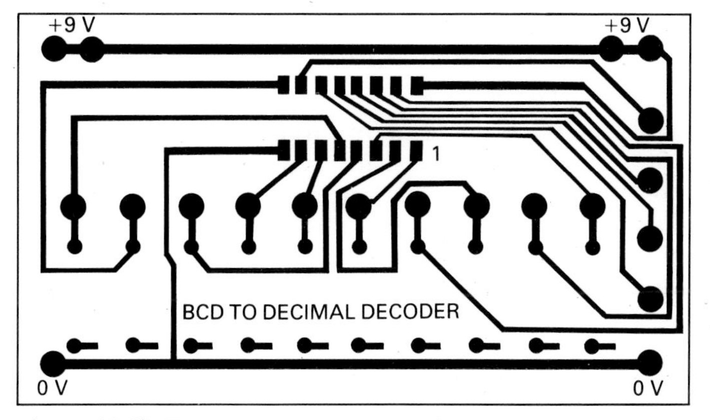

Figure 18.17 Track pattern on the PCB (actual size)

18.7 *Project Module* E6

Infrared Remote Control

What it does

These two Project Modules, a transmitter and a receiver, enable devices to be controlled remotely using an infrared beam. The output from the receiver is a 4-bit binary codeword which can be decoded further into 15 channels of control. The codewords are selected using a keypad on the transmitter unit. The output signals from the receiver can be used to energise relays for on/off power control of motors, lamps, etc. The Infrared Source and Sensor (Project Module B6) are used with the transmitter and receiver modules.

Circuits

The transmitter circuit is shown in figure 18.18 and is based on a single remote control transmitter, IC_1, a 490 type. This IC is able to modulate an ultrasonic beam if required, but here it modulates the infrared radiation emitted by LED_1. A pulse-position modulated

(PPM) beam of infrared radiation is generated when any one of the nine push switches, SW_1 to SW_9, is pressed. Each switch short-circuits a pair of pins on IC_1, e.g. pressing SW_1 short-circuits pins 1 (0 V) and 5, and a code is produced. The pulses generated by IC_1 switch transistors Tr_1 and Tr_2 on and off and provide the power to LED_1. The stream of pulses from LED_1 is focused into a parallel beam by the lens in the Infrared Source (Project Module B6) for transmission to the receiver.

The receiver circuit is shown in figure 18.19 and is based on a remote control receiver, IC_1, a 486 type. This IC amplifies the pulse-position modulated (PPM) signals picked up by the Infrared Sensor (Project Module B6), which incorporates a lens to collect and focus the infrared beam. The amplified PPM signal is delivered to the decoder, IC_2. This device responds to each of the 16 codewords transmitted by the 490 transmitter. Note that IC_2 operates on a supply voltage between 12 V and 18 V. IC_2 decodes the codeword which is latched on pins 5 to 8, i.e. the codeword is memorised by IC_2. Since the binary signals on these pins are low for logic 1 in the PPM codeword, IC_3 is used to invert the logic signals. LED_1 to LED_4 at the outputs of the NAND gates show which 4-bit codeword has been transmitted. If IC_2 is replaced by the 926 decoder, the signals are decoded but not latched, so the LEDs go out when the codeword stops being received. The variable resistor, VR_1, is used to tune the decoder to the frequency at which codewords are transmitted as explained below.

Components and materials

Transmitter

IC_1: remote control transmitter, type 490
Tr_1: pnp transistor, type BC479
Tr_2: npn transistor, type TIP31
SW_1 to SW_9: miniature push-to-make, release-to-break PCB switches
R_1 to R_4: fixed-value resistors, 0.25 W ± 5%, values 10 kΩ, 82 Ω, 33 kΩ and 2.2 kΩ, respectively

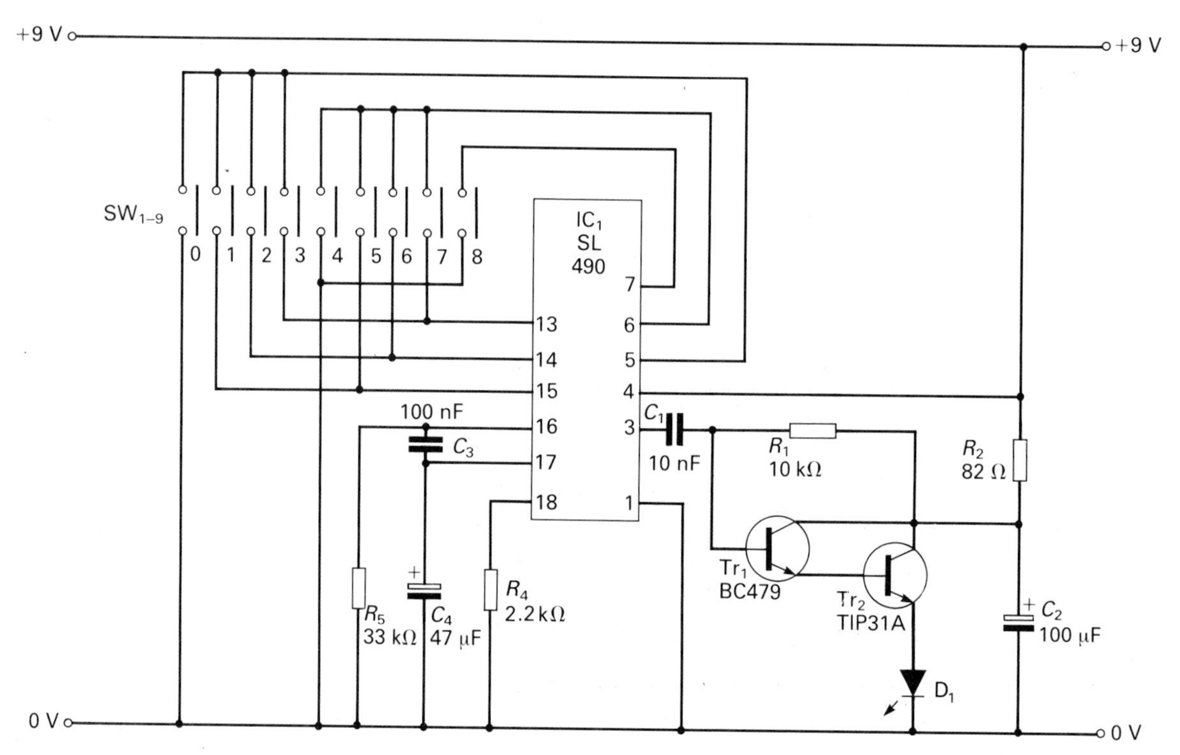

Figure 18.18 Circuit diagram of the Infrared Source

C_1, C_2: electrolytic capacitors, 100 μF and 47 μF, respectively
C_3, C_4: polyester capacitors, 100 nF and 10 nF, respectively
LED_1: infrared light emitting diode, type OP160, mounted in Infrared Emitter (Project Module B6)
IC holder: 1 × 18-way
battery: PP9
battery clip: PP9 type
wire: single strand, e.g. 7 × 0.2 mm
PCB: 90 mm × 50 mm
connectors: PCB header and PCB socket housing; crimp connectors

Receiver
IC_1: remote control preamplifier, type 486
IC_2: receiver, 16 latched 4-bit outputs, type 928
IC_3: quad 2-input NAND gate, CMOS type 4011
IC holders: 1 × 8-way, 1 × 14-way and 1 × 16-way
LED_1 to LED_4: red light emitting diodes
D_1: Infrared Sensor type OP500 (use Project Module B6)
R_1 to R_{12}: fixed-value resistors, 0.25 W, ± 5%, values 68 Ω, 47 Ω, 220 Ω, 5.6 kΩ, 4 × 3.3 kΩ, 4 × 2.2 kΩ, respectively
C_1, C_2 and C_7: electrolytic capacitors, values 10 μF, 47 μF, and 22 μF, respectively, 25 V working
C_3 to C_6, and C_8: polyester capacitors, values 47 nF, 4.7 nF, 220 nF, 220 nF, 22 nF, respectively
VR_1: miniature horizontal carbon preset resistor, value 100 kΩ

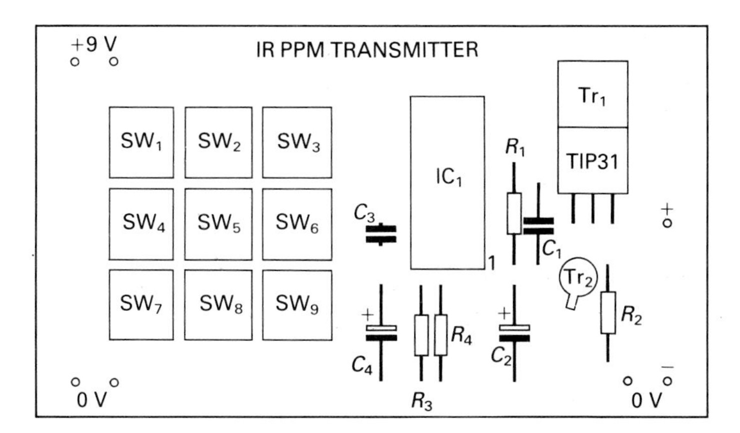

Figure 18.19 Circuit diagram of the Infrared Receiver

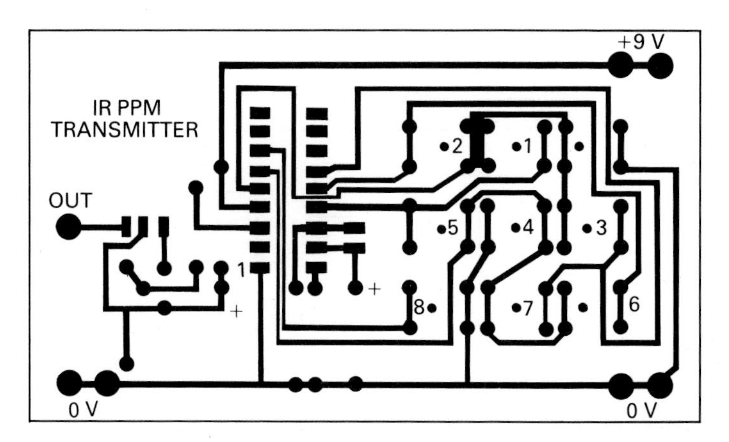

Figure 18.20 Component layout on the PCB (actual size)

battery: 2 × PP9
battery clip: 2 × PP9 type
wire: single strand, e.g. 7 × 0.2 mm
PCB: 90 mm × 50 mm
connectors: PCB header and PCB socket housing; crimp connectors

PCB assembly

Transmitter
Figure 18.20 shows the layout on PCB of the components for the transmitter, and figure 18.21 the copper track pattern on the other side of the PCB. See Book A, Section 6.3, for guidance on the preparation of the PCB.

Receiver
Figure 18.22 shows the layout on PCB of the components for the receiver, and figure 18.23 the copper track pattern on the other side of the board.

The terminal pins for the connections on the two PCBs are made by cutting single and double pins from the PCB header. Single sections of the PCB socket housing are used for terminating wire ends for connecting the infrared remote control to the battery and to other Project Modules. Strip 5 mm of insulation from the ends of the wires, and use a crimping tool to squeeze a crimp connector on the bare ends. Push the crimp connector into the PCB socket until it clicks into place.

Testing and use

Connect a PP6 (or PP3) battery to the transmitter. Note that, if you intend to house the transmitter, the power supply does not need an on/off switch since it draws current only during the intermittent use of the push-button switches, SW_1 to SW_9. An oscilloscope connected across

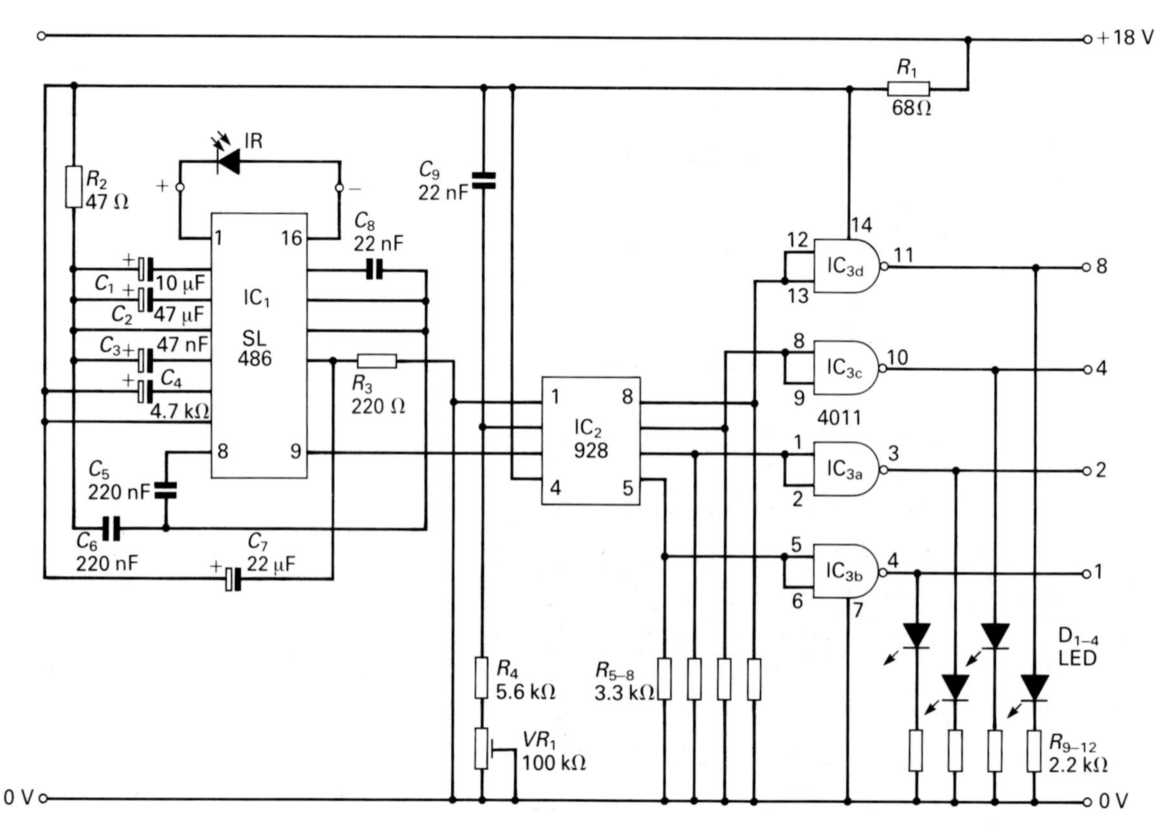

Figure 18.21 Track pattern on the PCB (actual size)

pin 3 and 0 V of the transmitter will show the patterns of the PPM codewords generated by I.C_1 when the switches are pressed. Long spaces between pulses represent binary 1 and short spaces binary 0. Note that this is a 5-bit codeword since IC_1 allows up to 32 codewords to be generated, though we have used only 16 of them. Switches SW_1 to SW_8 generate the first 8 codewords, 00001 to 01000. SW_9 sends out the codeword 00000 and delatches the decoder, IC_2, in the receiver.

By pressing SW_1 and SW_9 together, you can produce the codeword 01001, SW_2 and SW_8, 010010, and so on up to the codeword 01111 when SW_7 and SW_8 are pressed. Finally check that the codewords are present at the output pin of the transmitter; at this pin they appear as spiky patterns on the oscilloscope, but with the spikes still representing the coded pattern. Connect the Infrared Source (Project Module B6) across the output and 0 V lines of the PPM transmitter and a beam of invisible (and harmless) coded infrared radiation can be projected across a room.

Connect two PP9 battery connectors in series to provide an 18 V power supply for the PPM receiver. Connect the Infrared Sensor (Project Module B6) to the input of the receiver using only the diode, not the resistor in series with it in the Infrared Sensor module. Line up the two lens systems about a metre apart. Keep SW_1 pressed while you adjust VR_1 on the receiver until LED_1 lights to show the codeword 00001 has been received and latched. Press SW_9 and LED_1 will go out. Check that the other codewords can be latched and unlatched by the PPM receiver. Carefully focus the lens systems

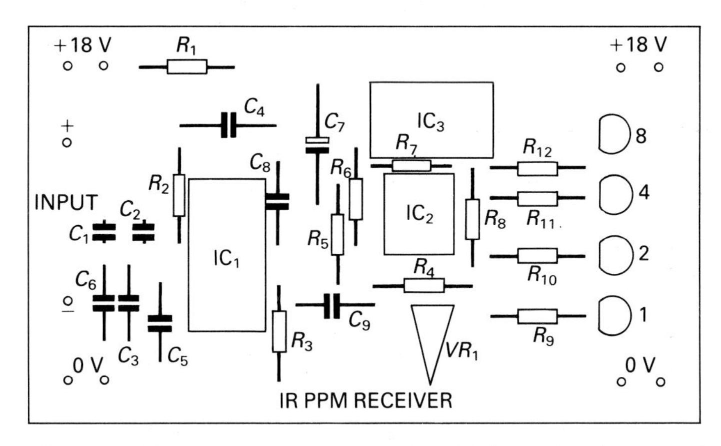

Figure 18.22 Component layout on the PCB (actual size)

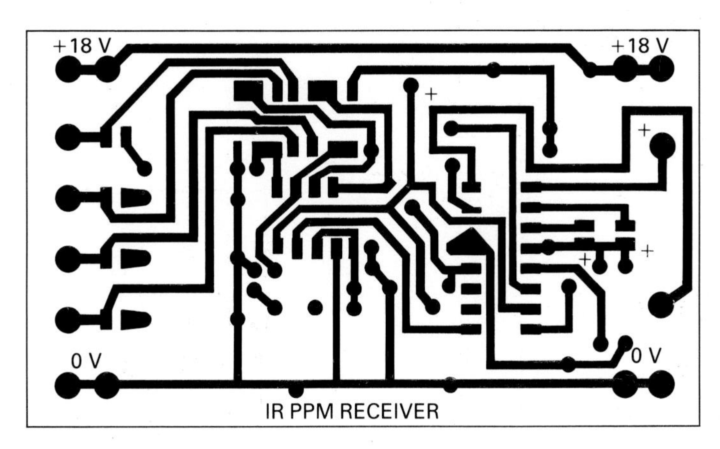

Figure 18.23 Track pattern on the PCB (actual size)

and check that the codewords can be received across a room.

Use two bits of logic high signals of the 4-bit binary code at the output of the PPM receiver to energise the two relays in the Relay Driver (Project Module B2).

Question

1 How would you use the Infrared Remote Control system to steer a simple vehicle powered by two d.c. motors?

Section 18.6 describes the design of a BCD-to-Decimal Decoder for converting the binary output from the Infrared Receiver into ten discrete channels. The CMOS 4028 device used in this decoder is only able to decode ten channels, not the full sixteen channels available. However, ten channels should be enough for most applications.

18.8 *Project Module* E7

64-bit Memory

What it does

This Project Module stores sixteen 4-bit words. Each word (called data) is held in an 'address' in the memory and this data can be read from the sixteen addresses in any order at random, i.e. the memory is a random access memory (RAM). The storage of data like this is the basis of computer memory systems. The data stored in the memory of this Project Module can be used to step devices through a controlled sequence of operations as explained below.

Circuit

The circuit shown in figure 18.25 is based on the 16 × 4-bit memory, IC_1. Simple memory devices of this type are discussed in Chapter 15. The four inputs, D_0 to D_3, are used for loading data into IC_1, and the four inputs, A to D, are used for specifying the addresses where this data is held. IC_2 inverts the data fed to IC_1 to ensure that the four LEDs, LED_1 to LED_4, indicate bit 1 when an LED is on, and a bit 0 when an LED is off.

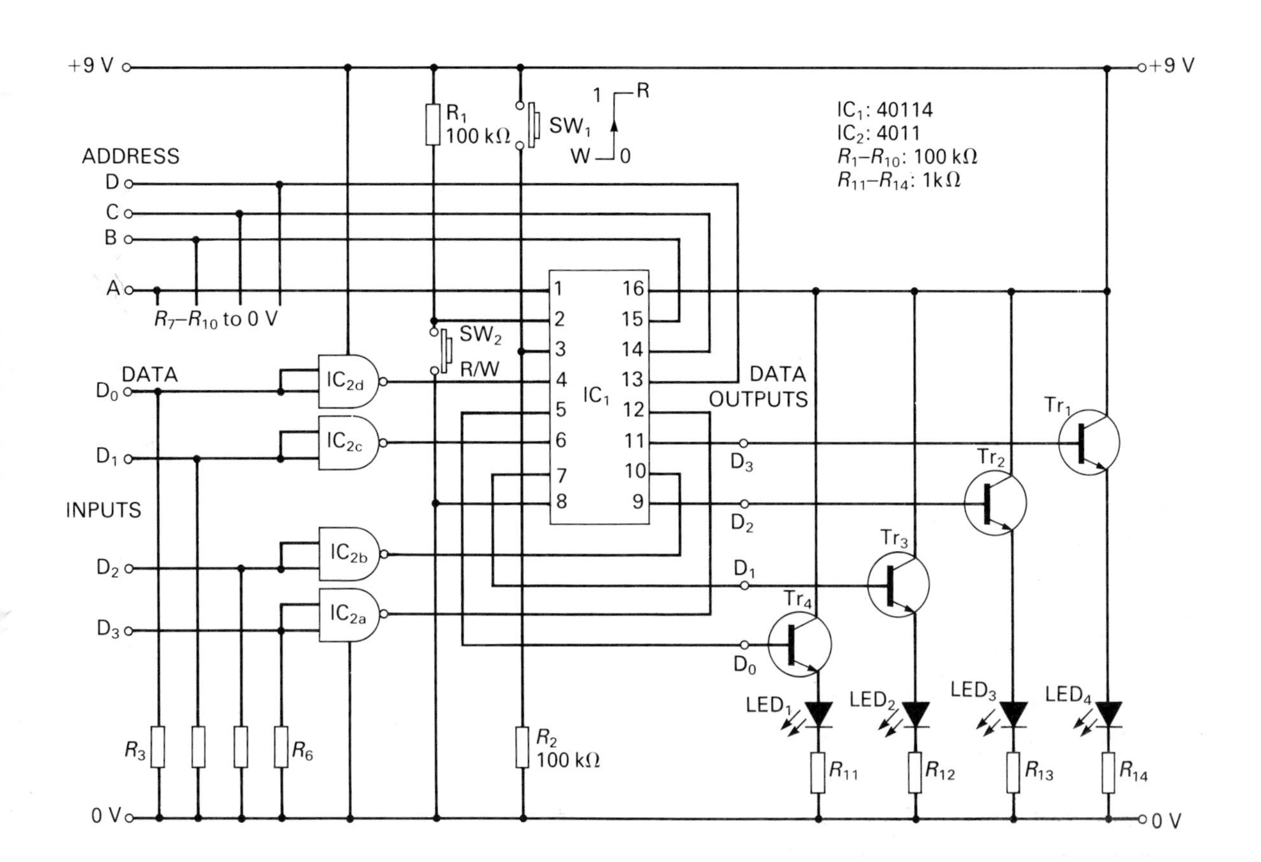

Figure 18.25 Circuit diagram of the 64-bit Memory

Components and materials

IC_1: 16 × 4-bit (64-bit) memory, CMOS type 40114
IC_2: quad 2-input NAND gate, CMOS type 4011
Tr_1 to Tr_4: npn transistors, type BC108
IC holders: 1 × 16-way, 1 × 14-way
SW_1: push-to-make, release-to-break keyboard switch
SW_2: 2-pole, 2-way miniature slide switch
LED_1 to LED_4: red light emitting diodes
R_1 to R_{16}: fixed-value resistors, 0.25 W ± 5%, values 10 × 100 kΩ and 4 × 1 kΩ
PCB: 90 mm × 50 mm
wire: multistrand, e.g. 7 × 0.2 mm
connectors: PCB header and PCB socket housing; crimp terminals

PCB assembly

Figure 18.26 shows the layout of the components on the PCB, and figure 18.27 the copper track pattern on the other side of the PCB. See Book A, Section 6.3, for guidance on the preparation of the PCB.

The terminal pins for the connections on the PCB are made by cutting single and double pins from the PCB header. Single Sections of the PCB socket housing are used for terminating wire ends for

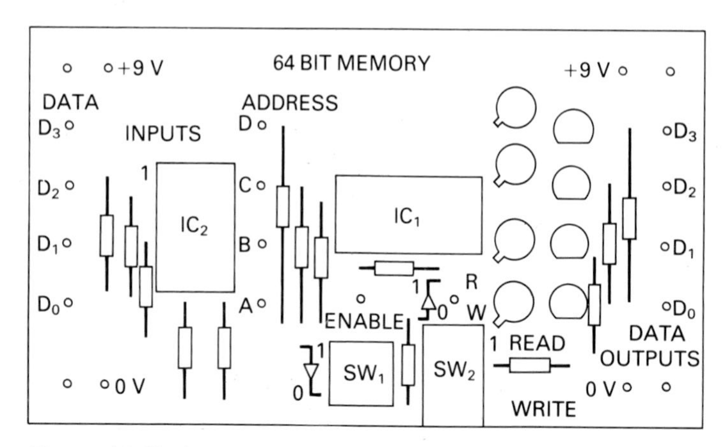

Figure 18.26 Component layout on the PCB (actual size)

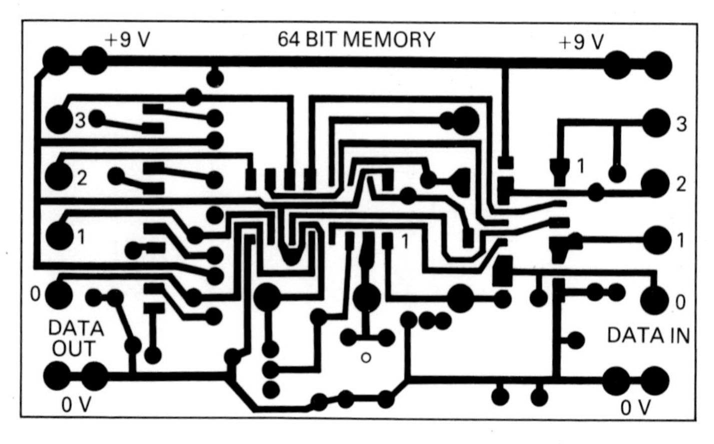

Figure 18.27 Track pattern on the PCB (actual size)

connecting the 64-bit Memory to the battery and to other Project Modules. Strip 5 mm of insulation from the ends of the wires, and use a crimping tool to squeeze a crimp connector on the bare ends. Push the crimp connector into the PCB socket until it clicks into place. Multiway connectors for the data and address lines can be made from 10-way lengths of connector socket. A crimp terminal is inserted into every other socket of each connector. Two such connectors are then joined by four 150 mm lengths of stranded wire.

Testing and use

Figure 18.28 shows the system required for loading data into the 64-bit Memory. Two Keyboard Encoders are required (Project Module E3), one of which provides sixteen 4-bit data words, and the other sixteen 4-bit addresses. To load data into the 64-bit memory proceed as follows:

(a) Set the READ/WRITE switch, SW_2, to WRITE.

(b) Press SW_7 (say) on Keyboard Encoder A so that the data presented to the 64-bit Memory is binary 0111. The LEDs will indicate this value

(c) Press SW_2 (say) on Keyboard Encoder B so that the address where this data is to go is 0010. the LEDs will indicate this value

(d) Press the ENABLE switch, SW_1, on the 64-bit Memory and the data 0111 is written into address 0010. This data cannot be seen on the four LEDs until the memory is read.

(e) Now set the READ/WRITE switch, SW_2, to READ.

(f) Press SW_2 on the Keyboard Encoder B, and the data 0111 will appear on the output LEDs of the 64-bit Memory.

(g) Check that data can be written and read from other addresses of the memory. For the moment, do not load data into addresses exceeding binary 1001. Do not forget to set SW_2 to WRITE when loading data into memory, and set it to READ when reading data from memory.

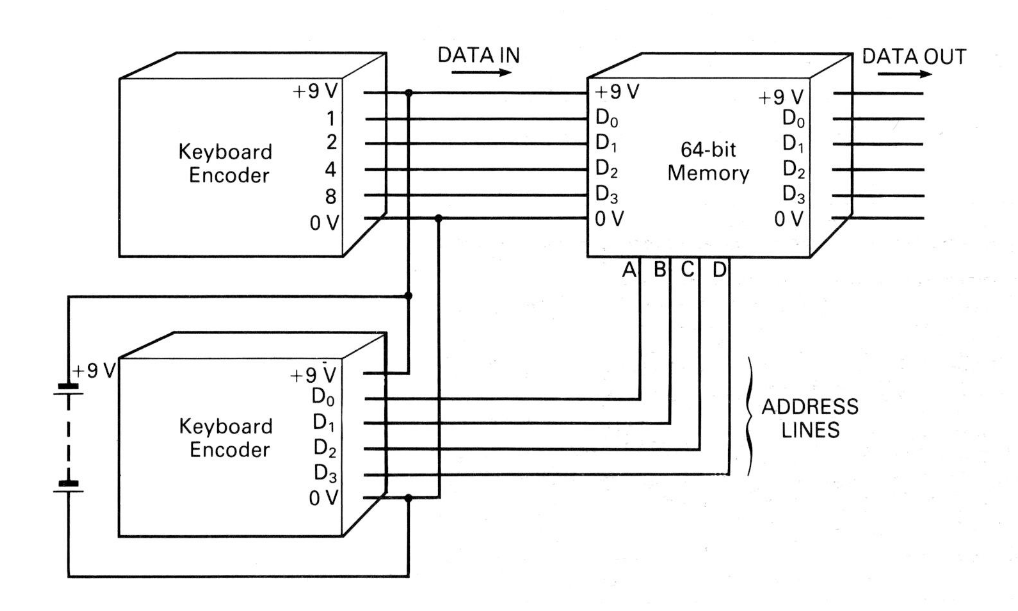

Figure 18.28 System for loading data into the 64- bit Memory

Figure 18.29 shows the system required for reading data from memory automatically. Make sure that the first ten addresses, i.e. 0000 to 1001, hold data that you have made a note of. The two Keyboard Encoders can be dispensed with, but four other Project Modules are required. Make sure you go through the following procedure:

(a) Choose appropriate values of resistor and capacitor to make the Pulser (Project Module A2) pulse at intervals of between 2 and 5 s. These output pulses operate the BCD Counter (Project Module E2). The Pulser also switches the Logic Gates (Project Module E1).

(b) Plug a 4011 NAND gate into the Logic Gates. One of the NAND gates is wired so that it operates as an inverter by joining together two inputs of this gate and connecting it to the output of the pulser. The output of the NAND gate is connected to the ENABLE pin on the 64-bit Memory.

(c) Set the BCD Counter to zero.

(d) Set the READ/WRITE switch, SW_2, to READ.

(e) Connect the Pulser to the input of the BCD Counter.

The LEDs on the BCD Counter will show the addresses cycling through from binary 0000 to binary 1001. Immediately after an address is presented to the 64-bit Memory, the output of the inverter goes from high to low and this 'enables' the memory so that the data stored in this address appears at the output of the 64-bit Memory.

Questions

1 How would you use the BCD-to-Decimal Decoder to convert the stored data into one of ten discrete signals?

2 Load data into the memory so that you can simulate a traffic lights sequence.

3 How would you adapt the BCD Counter module so that it provides access to all sixteen addresses in the 64-bit Memory?

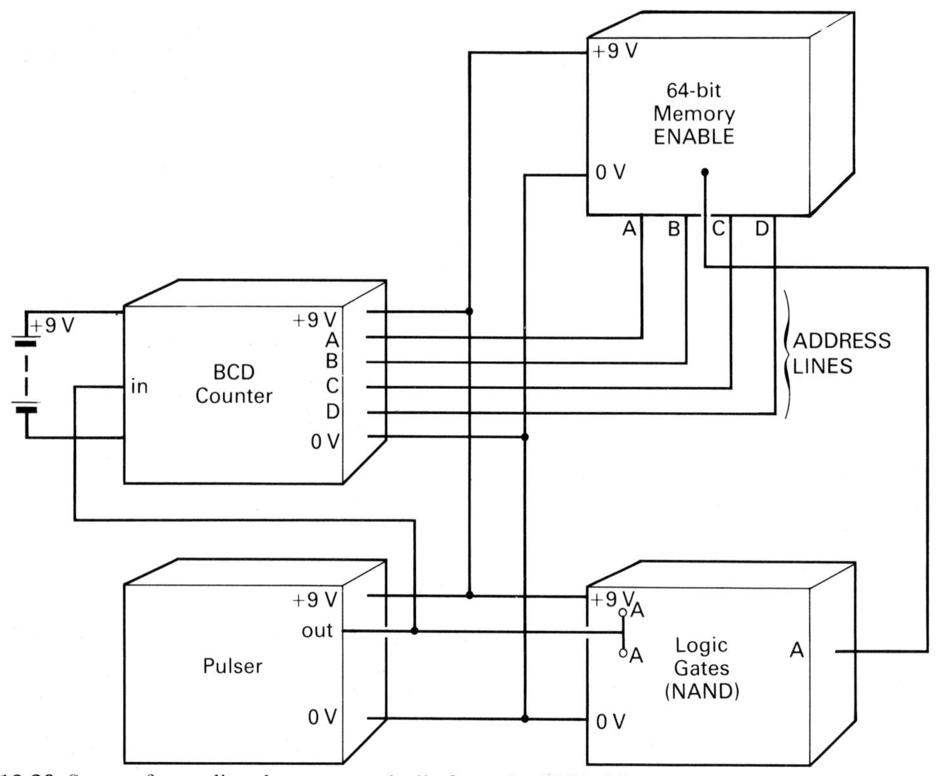

Figure 18.29 System for reading data automatically from the 64-bit Memory

19 Questions and answers – Book E

Revision questions

General ideas

1 A 'people counter' is installed at the entrance to an exhibition. Is the number of people counted an analogue or digital quantity?

2 'Gale force winds batter the lighthouse!' Is the speed of the wind an analogue or digital quantity? What is the name of an instrument for measuring windspeed?

3 Draw graphs to show the shape of analogue and digital signals.

4 The reed switch produces digital signals.
True or false?

5 The circumference of a bicycle wheel is 2 m and it is being ridden at 8 $m s^{-1}$. What is the speed of rotation of the wheel in revolutions per minute?

Displays

6 What is an alphanumeric display?

7 State three ways of displaying alphanumerics.

8 State two advantages of a liquid crystal display compared with a light emitting diode display.

9 What is the value of the resistor which you must connect in series with an LED to limit the current through it to 10 mA? Assume the voltage across the LED is 2 V and it is to be operated from a 12 V supply.

10 Draw a seven-segment display and label the segments. What numbers and letters can you obtain from this display?

Logic gates

11 What is the minimum number of inputs that an AND gate can have?

12 Two on/off switches connected in parallel act as . . .
An AND gate, an OR gate or a NOT gate

13 An OR gate's output is 1 when . . . .
all inputs are 0, only one input is 1

14 Write out the truth table for a 3-input NOR gate.

15 Write down the truth table for a NOT gate.

16 A NAND gate is called a universal logic gate because . . .
it turns up everywhere, it can be used to produce all other basic logic functions or it takes a lot of words to describe it?

17 Compare the supply voltage and power consumption of TTL and CMOS devices.

Logic problems

18 Show that the logic circuit in figure 19.1 functions as an exclusive-OR gate.

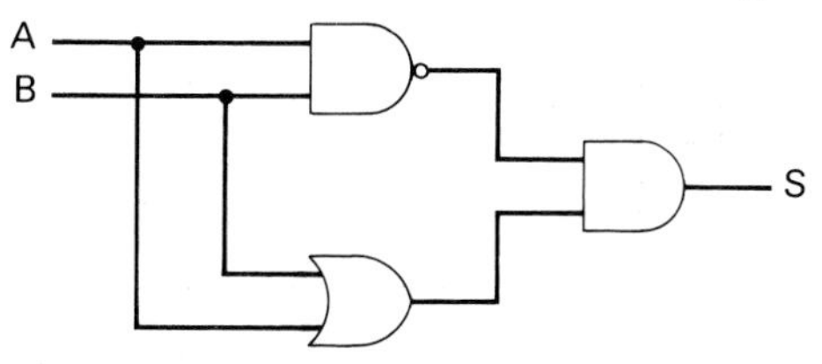

Figure 19.1

19 How many pins are there on an IC package which contains four 2-input NAND gates? Explain your answer.

20 Show a NAND gate can be used as a NOT gate.

21 How many 2-input NAND gates must be used to produce the 2-input NOR function?

Boolean algebra

22 Show that the arrangement of two 2-input AND gates in figure 19.2 is equivalent to a single 3-input AND gate.

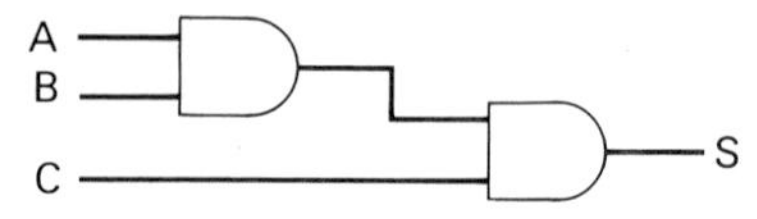

Figure 19.2

23 Write down the truth table for the logic diagram in figure 19.3.

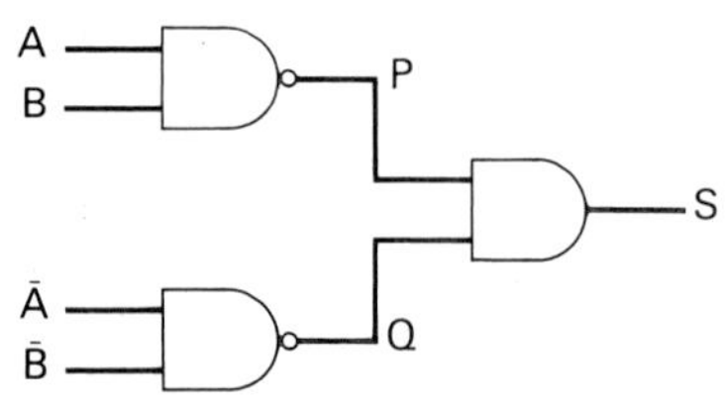

Figure 19.3

24 Write down the truth tables for the logic circuits in figure 19.4.

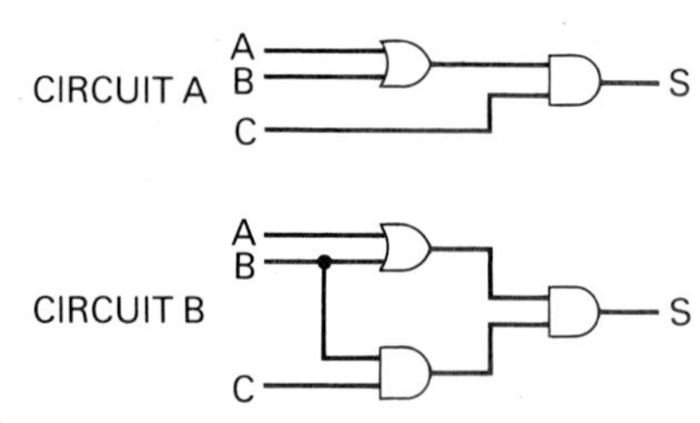

Figure 19.4

25 Draw a logic circuit that will allow a car to be started only if the gear lever is in neutral, the hand brake is on, and the seat belt is buckled. You have 2-input logic gates available for the design.

26 A processing plant uses four large tanks, A, B, C and D that contain different liquids. Tanks A and B have liquid level sensors, and tanks C and D have temperature sensors fitted to them. Assume that the level sensors are logic high when the level is too high, and logic low when the level is acceptable. Also assume that the temperature sensors indicate logic high when the temperature is too low, and logic low when the temperature is acceptable.

Design a logic circuit that provides an indication that the level in tanks A or B is too high at the same time as the temperature in tank C or D is too low.

27 A petrol store has four stations at each of which a tanker can be filled. But not more than three tankers are allowed to use the petrol store at any one time. Design a logic circuit that provides a warning signal if any three of the stations are used at any one time.

28 A piece of equipment has four fault detectors that signal a fault to an alarm circuit made up of logic gates. Make up a truth table that shows when any two detectors signal a fault, and design the logic circuit that would meet the requirement.

29 What do the following Boolean expressions mean?

$$\overline{A} = S; \; A.B + \overline{C} = S; \; A + \overline{B + C} = S$$

30 Design a logic system which will switch on one of the lamps L_1, L_2 or L_3 in figure 19.5 to indicate the height (high, medium or low) of each one of the lorries. Assume that the breaking of a beam of light by a lorry passing in front of the sources S_1 to S_3 produces a logic 1 input to the logic system.

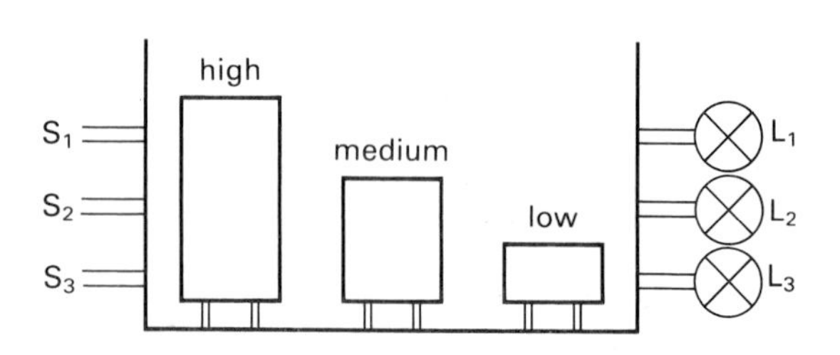

Figure 19.5

31 Use truth tables to prove the following relationships are correct:
(a) $A.(A + B) = A$
(b) $A + \overline{A} = 1$
(c) $(A + B).(A + C) = A + (B.C)$
(d) $\overline{A + B + C} = \overline{A}.\overline{B}.\overline{C}$

32 Simplify the following Boolean expression:

$$A.\overline{B}.\overline{C} + A.B.\overline{C} + \overline{A}.\overline{C}$$

Flip-flops and counters

33 The two outputs of a flip-flop are known as complementary outputs because . . .
in use they warm up, they always have opposite logic states or they act in unison

34 List two advantages of CMOS ICs over TTL ICs.

35 What is a binary counter?

36 Draw a diagram to show how two flip-flops can be used as a 2-bit binary counter.

37 Four BCD counters are cascaded.
(a) What is the maximum decimal count possible?
(b) If the BCD counters are reset and 369 counts are delivered to the 'units' counter, what do the BCD counters read?

38 Draw a diagram to show how binary counters can be used to give a 1 Hz pulse from the mains alternating current supply.

39 (a) Explain the statement: 'a flip-flop is a divide-by-two counter'.
(b) Show by means of sketch how four flip-flops can be wired together to produce a 4-bit counter.

40 Given a BCD counter and two flip-flops show how you would convert a 1 MHz clock frequency into a 25 kHz clock frequency.

41 Design a 10 minute timer using a 1 Hz oscillator and IC counters to divide down the frequency.

42 A digital watch has a crystal clock that operates at a frequency of 32 kHz. Given flip-flops, 4-bit binary counters and decade counters how would you reduce this frequency to a 1 Hz clock pulse?

43 If a signal of frequency 512 kHz is input to seven cascaded flip-flops, what is the frequency of the signal that emerges from the seventh flip-flop?

44 The number of BCD counters required to count up to 99 999 is . . .
five, six or seven?

45 Figure 19.6 shows a 4-bit binary counter which is fed with a regular series of clock pulses. The AND gate selects two of the bits from the 4-bit counter and its output is fed to the reset terminal of the counter. Explain why the counter counts ten pulses before resetting to zero.

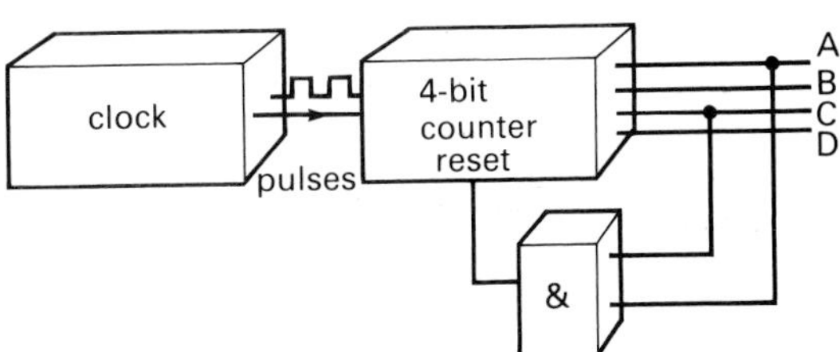

Figure 19.6

46 Figure 19.7 shows the layout of the seven segments of a seven-segment display. Each segment can be illuminated by applying a logic 1 to that segment. The ten decimal numbers 0 to 9 can be displayed by illuminating various combinations of these segments.

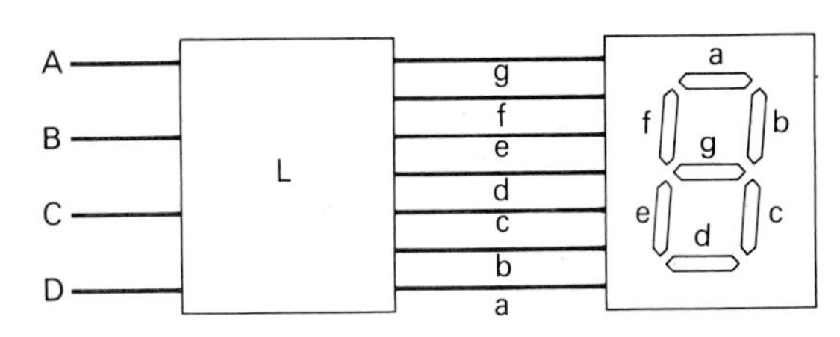

Figure 19.7

Design the logic circuit contained in block L which takes ten 4-bit numbers and produces the logic levels at its outputs to switch on the correct segments to display the ten decimal numbers.

47 Design a 10 Hz oscillator based on the CMOS 4093 IC, or on the TTL 7414 Schmitt trigger IC.

48 Describe three different ways of producing clock pulses for inputting to a digital system.

Number systems

49 What are the binary numbers equivalent to the following decimal numbers
(a) 7 (b) 27 (c) 64 (d) 255?

50 What are the BCD numbers for the following decimal numbers
(a) 17 (b) 99 (c) 201
(d) 1028?

51 What are the decimal numbers for the following BCD numbers
(a) 0010 (b) 1001 0010
(c) 0111 1000 0101

52 Decimal 15 equals . . . in hexadecimal.
A?/B1?/F?

53 Hexadecimal A6 equals . . . in decimal.
166?/190?/255?

54 What is the most important characteristic of the Gray code?

55 What are the Gray code numbers of the following decimal numbers.
(a) 6 (b) 9?

56 Add the following binary numbers.
(a) 110 + 10 (b) 111 + 101
(c) 1011 + 1101

57 Subtract the following binary numbers.
(a) 1011 − 10
(b) 10101 − 100111
(c) 11100001 − 10011110

Memory

58 What is the main difference between RAMs and ROMs?

59 What is the main difference between static RAMs and volatile RAMs?

60 Briefly describe the following terms used in association with memory:
bit, byte, data, read, write, volatile memory, random access, access time, memory cell.

61 Suppose a particular semiconductor memory can store 32 768 bits of data. What is its word capacity expressed in K if each word contains 8 bits?

62 A particular integrated circuit memory has a capacity of 64 words of 4 bits each. Calculate:
(a) the number of select lines on the IC;
(b) the number of data input and data output pins on the IC; and
(c) the number of memory cells there are on the chip inside the package.

Revision answers

5 $8\ \text{m s}^{-1}/2$ or 4 revs per second = 240 revs per minute.
9 Resistance = voltage across resistor/current through resistor = 10 V/10 mA = 1 kΩ
10 Numbers 0 to 9; letters b, c, d, h, l, n, o, t, u, A, B, C, E, F, H, L, O, P, S, U

14

A	B	C	S
0	0	0	1
0	0	1	0
0	1	0	0
0	1	1	0
1	0	0	0
1	0	1	0
1	1	0	0
1	1	1	0

19 14 pins
21 4
38 (a) 9999 (b) 0000 0011 0110 1001
43 10
44 2 kHz
45 (a) 5
47 (a) 0010 (P); 0111 (Q)
(b) $14.1\ \text{m s}^{-1}$

49 15 times
50 (a) 0111 (b) 11011 (c) 1111111
51 (a) 0001 0111
(b) 1001 1001
(c) 0010 0000 0001
(d) 0001 0000 0010 1000
52 (a) 2 (b) 92 (c) 785
53 F
54 166
56 (a) 0101 (b) 1101
57 (a) 1000 (b) 1100 (c) 11000
61 2048
62 1048576

Answers to questions

Section 3.3
1 Output 0 except when all inputs at 1, then 1.1.1 = 1
Section 3.4
1 Output 1 if one or more inputs at 1, i.e. 1 + 0 + 0 + 0 = 1; output 0 when all inputs at 0
Section 3.6
1 Output 1 except when all inputs at 1, then 1.1.1.1 = 0
Section 3.7
1 Output 0 except when all inputs at 0, then 0 + 0 + 0 + 0 = 1
Section 3.8

1

A	B	XOR	XNOR
0	0	0	1
0	1	1	0
1	0	1	0
1	1	0	1

(see figure 3.9)
Section 4.2
1 0
Section 4.3
1 (a) R = A OR B (A + B): S = (A OR B) AND C; thus S = (A + B).C
(b) Q = NOT A: R = NOT $\overline{A}$ AND B; S = (NOT A) AND B OR B;
thus S = ($\overline{A}$.B) + B

2 (a)

A	B	C	S
0	0	0	0
0	1	0	0
1	0	0	0
1	1	0	0
0	0	1	0
0	1	1	1
1	0	1	1
1	1	1	1

(b)

A	$\overline{A}$	B	S
0	1	0	0
0	1	1	1
1	0	0	0
1	0	1	1

3 (a)

A	B	$\overline{A}.\overline{B}$ = S
0	0	1
0	1	0
1	0	0
1	1	0

(b)

A	B	R	0 V	S = A + B + 0
0	0	1	0	0
0	1	0	0	1
1	0	0	0	1
1	1	0	0	1

4

A	B	C	A + B.C = S
0	0	0	0
0	0	1	0
0	1	0	0
0	1	1	1
1	0	0	1
1	0	1	1
1	1	0	1
1	1	1	1

5 $A = \overline{A}.B + A.\overline{B}$

A	B	S
0	0	0
0	1	1
1	0	1
1	1	0

This is the truth table for an exclusive-OR gate – see figure 3.9.

Section 5.6
1 16
2 (a) 3 (b) 4

Section 5.7
1 (a) 99 999 (b) 0000 0110 0100 0010 0101
2 (a) 32 000
(b) three cascaded BCD counters divide by 1000 giving 32 Hz; followed by
one 4-bit counter to divide by 16 giving 2 Hz;
followed by
one flip-flop to divide by 2 giving 1 Hz

Section 5.8
1 0110 0100 0000 0000 0000

Section 9.8
1 22

Section 10.1
1 $3 \times 10^3 + 6 \times 10^2 + 2 \times 10^1 + 1 \times 10^0 + 5 \times 10^{-1}$

Section 10.2
1 19.5

Section 10.3
1 1000011_2
2 1100011.111_2

Section 10.4
1 163

Section 10.5
1 52_{16}
2 $17.CC_{16}$

Section 10.6
1 E_{16}
2 479_{16}

Section 10.7
1 11010011_2
2 11111111_2

Section 10.8
1 10101_2 or 21_{10}
2 101_2
3 1111_2

Section 10.9
1 DA_{16} or 218_{10}
2 21_{16} or 33_{10}
3 $8E_{16}$ or 10001110_2

Section 11.2
1 (a) 0101 1100 (b) 1000 1001 0011
2 (a) 93 (b) 7806
3 (a) 1011 (b) 0100 0101 0100

Section 11.3
1 (a) 0010 (b) 1011
2 (a) 1000 (b) 1111

Section 12.3
1 Alarm = K.(W + D) = A

$$A = \overline{\overline{K.(W + D)}} = \overline{K} + \overline{(\overline{W + D})}$$

Logic circuit as below.

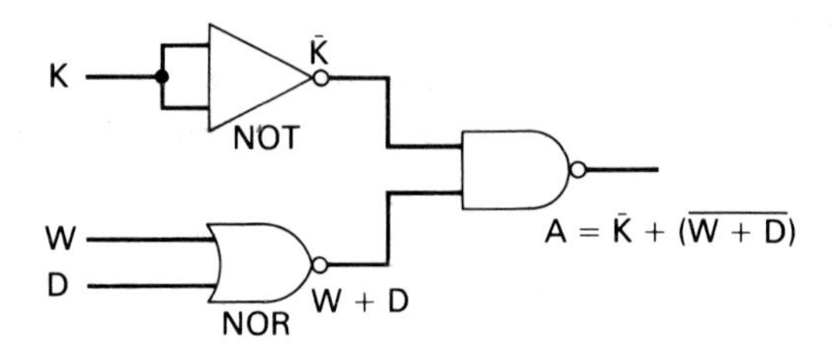

Figure 19.8

2 Open = M.C + D.C = O. Same solution as for Example 2 (Animal House) using NAND gates.

3 $\text{Alarm} = L = \overline{\overline{A.B + A.C + B.C}}$

$$= \overline{\overline{A.B}.\overline{A.C}.\overline{B.C}}$$

Logic circuit as below.

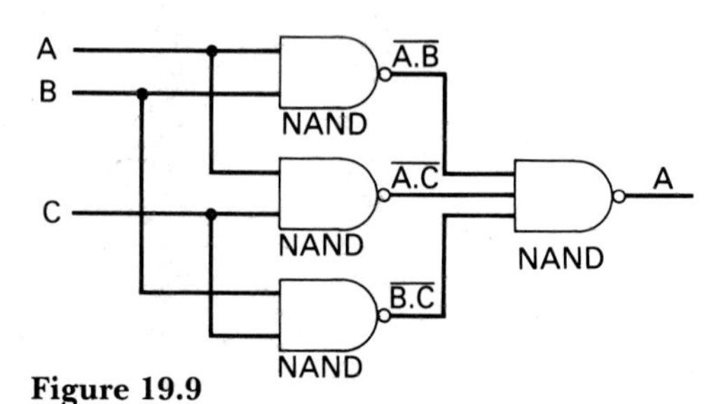

Figure 19.9